STATE OF DISASTER

State of Disaster

The Failure of U.S. Migration Policy
in an Age of Climate Change

→

MARIA CRISTINA GARCIA

The University of North Carolina Press
CHAPEL HILL

This book was published with the assistance of the Thornton H. Brooks Fund of the University of North Carolina Press.

© 2022 The University of North Carolina Press

All rights reserved
Set in Warnock Pro, ITC Franklin Gothic Std
by codeMantra

Manufactured in the United States of America

The University of North Carolina Press has been a member of the Green Press Initiative since 2003.

Cover photo: The U.S. Coast Guard transporting St. Thomas residents during Hurricane Irma, September 12, 2017. Photo courtesy of the U.S. Coast Guard.

LIBRARY OF CONGRESS CATALOGING-IN-PUBLICATION DATA
Names: García, María Cristina, 1960– author.
Title: State of Disaster: The Failure of U.S. Migration Policy in an Age of Climate Change / Maria Cristina Garcia.
Description: Chapel Hill : University of North Carolina Press, [2022] | Includes bibliographical references and index.
Identifiers: LCCN 2022007368 | ISBN 9781469669953 (cloth ; alk. paper) | ISBN 9781469669960 (paper ; alk. paper) | ISBN 9781469669977 (ebook)
Subjects: LCSH: Environmental refugees—Government policy—United States. | Disaster relief—Government policy—United States. | Disaster relief—Government policy—United States—Territories and possessions. | Environmental disasters—Social aspects—Caribbean Area—Case studies. | Environmental disasters—Social aspects—Central America—Case studies. | Climatic changes—Social aspects. | United States—Emigration and immigration—Government policy. | United States—Foreign relations—Caribbean Area. | United States—Foreign relations—Central America.
Classification: LCC HV640.4.U54 G38 2022
LC record available at https://lccn.loc.gov/2022007368

For Alexandra, Nicholas, Isabelle, Natalie, and Cristina

CONTENTS

LIST OF ILLUSTRATIONS • viii

ACKNOWLEDGMENTS • xi

INTRODUCTION. In Search of a Sustainable Refuge • 1

CHAPTER 1. What a Volcano Revealed about the Vulnerability of Small Island States • 19

CHAPTER 2. Disaster Relief as Foreign Policy: When Poverty, Conflict, and Catastrophe Collide • 47

CHAPTER 3. What Protections and Benefits? Coloniality and Citizenship in the U.S. Territories • 93

CONCLUSION. Moving Forward: Natural Disasters May Be Inevitable; Good U.S. Policy Is Not • 139

NOTES • 151

INDEX • 225

ILLUSTRATIONS

FIGURES

Sign warning of danger from volcano in the Montserrat exclusion zone • 39

Plymouth, the old capital of Montserrat, buried under mud and lava flows • 42

Ash and steam plume, Soufrière Hills volcano • 44

Soto Cano Air Base, Honduras, the hub of the international relief effort • 60

President Bill Clinton addresses the people of Posoltega, Nicaragua • 66

Humanitarian aid from Travis Air Force Base, California • 97

Donald Trump at Muñiz Air National Guard Base, Puerto Rico • 112

Protest against Governor Ricardo Rosselló, San Juan • 135

ILLUSTRATIONS

MAPS

Montserrat • 22

Honduras and Nicaragua • 50

The U.S. Virgin Islands and Puerto Rico • 94

ACKNOWLEDGMENTS

Writing about the recent past always poses methodological challenges, but writing about recent environmental disasters can be especially daunting. I often found it difficult to put the research aside at the end of the day. It affected how I viewed everything in my daily life, from weather forecasts and gardening to consumerism and civic engagement. The support of family and friends was essential. I will thank them later in the acknowledgments, but here—at the very beginning of the book—I want to offer my most heartfelt gratitude to my husband, Sherman Cochran, whose love and support has sustained me through the best and worst of times. Sherm read chapters, asked great questions, and listened to my concerns, but more importantly, he helped me laugh and be hopeful. I could not have taken on this project without him.

Finding time for research and writing has become more and more elusive over the course of my career. Thanks go to the Carnegie Corporation of New York for the fellowship that temporarily released me from my teaching and administrative responsibilities so I could give careful thought to this project and immerse myself in the multidisciplinary scholarship on climate change. I am honored to have been a 2016 Andrew Carnegie Fellow, and I thank the colleagues at Cornell University who nominated me for this distinguished fellowship.

ACKNOWLEDGMENTS

I am grateful to the librarians, archivists, and specialists who answered my many questions or located information for me at the Migration Statistics Unit of the UK Office of National Statistics, the National Archives at Kew (UK), the U.S. National Archives and Records Administration, and the George H. W. Bush and William J. Clinton Presidential Libraries. I am especially grateful to archivists R. Matthew Lee and Michelle Bogart at the Bush Presidential Library and Jason Kaplan, John Keller, and Herbert Ragan at the Clinton Presidential Library. The digitized records of the Center for Puerto Rican Studies at Hunter College, the Benson Latin American Collection and Teresa Lozano Long Institute for Latin American Studies at the University of Texas at Austin, and the National Security Archive, among other institutions, were essential, especially during 2020 and 2021, when physical travel to archives became impossible due to the COVID-19 pandemic. At Cornell, my home institution, the librarians of Olin, Uris, Mann, and Law Libraries were always helpful. I owe special thanks to Sarah Kennedy, who continually alerted me to sources that might be helpful to my project and then seemingly fast-tracked their availability.

I presented work in progress at various academic and public settings, and the questions and comments from the audience helped me fine-tune my ideas and make this a stronger book. I am grateful to the students, faculty, and staff at Gustavus Adolphus College, where I gave the Raoul Wallenberg Memorial Lecture in Peace Studies; the Institute for Global Studies and the Immigration History Research Center at the University of Minnesota; the Julie Ann Wrigley Global Institute of Sustainability at Arizona State University; American University, where I gave the Gary Braithwaite Memorial Lecture; Rice University, where I gave the Ervin Frederick Kalb Annual Lecture; the Institute for the Humanities at the University of Illinois, Chicago; and the Kahn Liberal Arts Institute at Smith College. Thanks go to the colleagues who hosted me and to the staff who arranged the details of my visits (in alphabetical order): Alexander Aviña, Seán Easton, Emir Estrada, Adam Goodman, Sayuri Shimizu Guthrie, Meghan Harrick, Paul Hirt, Dirk Hoerder, Alan Kraut, Erika Lee, Julian Lim, Maddalena Marinari, Bryan Messerly, Saengmany Ratsabout, and Eric Weld.

At Cornell, I presented work in progress at the Judith Reppy Institute for Peace and Conflict Studies, the Atkinson Center for Sustainability, the Department of Natural Resources, the history department's Comparative History Colloquium, the Environment and Sustainability Colloquium, the Cornell Botanic Gardens, the Migrations Initiative, The Circle, the Africana Studies and Research Center, the Cornell Association of Professors

ACKNOWLEDGMENTS

Emeriti, and the Hans Bethe Residential College. Many thanks are due the colleagues who invited me and helped facilitate lively discussions: Shorna Allred, Christopher Dunn, Matt Evangelista, Shannon Gleeson, Peter Hess, Johannes Lehmann, Susan Jean Riha, Eric Tagliacozzo, Julia Thom-Levy, Wendy Wolford, and Steven Yale-Loehr.

I am grateful for the faculty and graduate student colleagues and friends who read parts or all of the chapters or gave feedback during long discussions over food and drink: Spencer Beswick, Laura Brown, Judith Byfield, Vicki Caron, Sherman Cochran, Claire Cororaton, Ray Craib, James Cutting, Tim Devoogd, Emily Estelle, Howard Feinstein, David Feldshuh, Kaitlin Findlay, Jason Frank, Jill Frank, Kevin Gaines, Durba Ghosh, Larry Glickman, Lyrianne González, Sandra Greene, Isabel Hull, Karen Jaime, Carol Kammen, Peter Katzenstein, Claudia Lazzaro, David Lodge, Veronica Martínez-Matsuda, Peter McClelland, Sarah Meiners, Mary Beth Norton, Marcos Pérez Cañizares, Jeremy Peschard, Aaron Sachs, Daniela Samur, Susana Romero Sánchez, Josh Savala, Bob Stoll, Barry Strauss, Robert Travers, and Penny Von Eschen.

The staffs of the history department, the Latina/o Studies Program, and the American Studies Program have offered friendship and support in countless ways over the years: Adrienne Clay, Marti Dense, Barb Donnell, Katie Kristof, Azucena Ortega, Claire Perez, Georgiana Saroka, Kay Stickane, Michael Williamson, and Judy Yonkin. What good fortune I've had to have Barry Strauss and Tamara Loos as history department chairs during this period; I am grateful for their enthusiasm and collegiality. In Latina/o studies, where I hold a joint appointment, and American studies, where I am an affiliated faculty member, Debra Castillo, Vilma Santiago-Irizarry, and Noliwe Rooks were supportive program directors.

My undergraduate research assistants, Isabelle Maria Kiefaber and Leighton Fernando Cook, helped me track down Spanish language sources. (Isabelle and Leighton, I apologize if the detective work and collection of data discouraged you from ever considering careers as historians.) My dear friends Harold Livesay and Cynthia Bouton welcomed me to their College Station home on numerous occasions so I could have prolonged access to the Bush Presidential Library. Antonio and Carol Pérez hosted me during my many trips to the Washington, D.C., area to consult sources and conduct interviews. My friends Julia and Jere Blackwelder offered lively conversation and excellent meals at their home.

I had the exceptional good fortune to work with Elaine Maisner at the University of North Carolina Press. Elaine believed in this project from the start and discussed my ideas at length by Zoom, phone, and email. She helped make this a better book, and I am so grateful to her. I also offer my sincerest thanks to Andreina Fernandez for answering my many tedious questions and to Mark Simpson-Vos, who met with me at conferences when Elaine was unable to attend. I am grateful for the assistance of the editorial, production, and marketing teams at UNC Press, especially Julie Bush, Erin Granville, Elizabeth Orange, and Dino Battista.

The anonymous readers were among the best I've worked with over the past thirty years. I thank them for their careful reading of the chapters and their excellent suggestions. I hope they are pleased with the results.

During the writing of this book, I completed a three-year term as president of the Immigration and Ethnic History Society. I want to thank my colleagues Madeline Y. Hsu, Timothy Draper, Cheryl Greenberg, Barbara Posadas, Hasia Diner, Maddalena Marinari, and Kevin Kenny. I could not have taken on this professional service without their enthusiastic support and assistance. Together we accomplished a great deal. I look forward to the changes that are yet to come, as a new generation of migration scholars takes this professional association in interesting new directions.

Several colleagues passed away while I wrote this book, and I want to acknowledge their friendship and the role they played in my development as a scholar: Harold Livesay and Larry Hill at Texas A&M University, where I taught for the first nine years of my career, and Michael Kammen, Dick Polenberg, Walt LaFeber, Joel Silbey, Bob Smith, and Kazuko Smith at Cornell, where I have taught since 1999. All of them were brilliant scholars, but more importantly they were kind and decent human beings who cared deeply about the world. I will treasure the memories of our many conversations about history, sports, politics, and the arts. As this book goes to press, a longtime friend, Dean Rich, has passed away unexpectedly. I will miss his sharp questions, his quick wit, and the ways he pushed me and other academics to be accountable to readers outside of academia.

I began the acknowledgments by thanking my husband, and I will end it by thanking other members of my family who have been equally important. A generation ago, in a different country, my multigenerational extended family lived within walking distance of each other, which made it easy to be part of each other's lives. We now reside in the United States, dispersed across our adopted country, yet the love and unconditional support endure. My mother is now the oldest matriarch of the family. We talk every day by phone, sometimes for hours, and her faith and courage always lift my spirits. (Yes, I know how fortunate I am.) In Texas, my sister, Victoria, and brother-in-law, Edward, inspire me with their optimism and sense of adventure. In Wyoming, my brother, Joe, and my sister-in-law, Renee, have offered support through thick and thin. In Virginia, my cousin Tony and his wife, Carol, are two of the funniest and sharpest people I know. In California, my cousin Alessandra inspires me by her commitment to family, her perseverance through challenging times, and her dedication to her middle school students. In Florida and Puerto Rico, countless cousins have made life so interesting. Family members have also passed away during this period, and their absence has left a hole impossible to fill. I carry the love of my aunts, Martha H. Pérez and Mirtha R. Vega, and my cousin Vivian Argilagos with me always. Their memory is a blessing.

This book is dedicated to my godchildren, Alexandra and Nicholas Pérez and Isabelle, Natalie, and Cristina Kiefaber. Allie, Nick, Izzy, Nina, and Crissy have come of age in an era of unanticipated challenges, but they have the intelligence, skills, creativity, courage, and heart to move confidently in the direction of their dreams. They inspire me every day.

STATE OF DISASTER

INTRODUCTION

In Search of a Sustainable Refuge

Historians have long been fascinated by the ways short-term and long-term developments in the weather have shaped exploration and settlement, agricultural development and trade, and military battles and war. Historians who study migration know that extreme weather events, such as hurricanes, earthquakes, and volcanic eruptions, and longer-term environmental degradation, such as soil erosion, drought, and desertification, have displaced people for millennia. It's part of the human story. These conditions have served as tipping points that have set societal change into motion or accelerated changes already underway. They have disrupted livelihoods and sparked sectarian conflict, destabilized governments, and forced people to migrate within countries and across imperial, continental, and national borders.

We are now living in an era of accelerating climate change.[1] At no other time in human history have we had 8 billion people living on the planet, all contributing in some way to the greenhouse gas emissions, pollution, deforestation, and other behaviors that are warming the planet and hastening changes that more commonly occurred in the past over thousands of years rather than decades.[2] The sustained rise in global temperatures has led to the unusually rapid melting of polar ice sheets, permafrost, and glaciers that is resulting in sea level rise, erosion, and the salination of soil and water. The higher temperatures are also shifting weather patterns, leading to more extreme, rapid-impact events such as hurricanes, typhoons, storm surges, and flooding in some parts of the world and drought, desertification, and wildfires in others. These disruptions have produced crop failures and fresh water shortages, infrastructure breakdown and energy blackouts, infectious disease outbreaks, and the loss of arable land, natural habitat, and wildlife.[3]

Humans have adapted to sudden-onset disasters and environmental change—and the societal disruptions that come with them—through migration, but our current era offers particular challenges. As environmental historian John R. McNeill has written, our Paleolithic ancestors coped with rapid climate change from time to time, but they did so when the earth had fewer people than Chicago has today. Their response to climate change was to move elsewhere; but "since the emergence of agriculture, sedentarism, civilization, and the settlement of all habitable parts of the globe, the Paleolithic response has become more impractical."[4] In an era of climate change, the disappearance of livelihoods and the threats to safety will require that tens of millions of people will have to relocate temporarily or permanently in the decades to come. "While there are analogues in Earth's history for the climate change now under way," wrote McNeill in 2007, "there are none in human history. We have entered uncharted terrain."[5]

Why These Cases?

This book examines U.S. responses to environmental disruptions in Central America and the Caribbean since 1995 to see what lessons might be learned for shaping humanitarian and immigration policies in an era of accelerating climate change. The environmental disasters discussed in this book cannot be attributed conclusively to anthropogenic climate change; compiled data and observable patterns may be suggestive of a correlation between human activity and a specific disaster but not conclusive.[6] The case studies were chosen for what they reveal about climate shocks, the internal and cross-border migration that follows, and the lack of legal protections for those displaced by environmental drivers and forced to migrate. They offer a view into some of the humanitarian challenges of the present and the future.

These chapters are not solely about migration, however. Disasters draw people into poverty, exacerbate inequality, and create multiple forms of vulnerability that lead to internal and cross-border migration, but disasters also reveal the failures in planning, development, and governance both before and after a crisis. They reveal the legacies of imperial and Cold War policies. Examining these failures and legacies are essential to understanding not only the context for migration but also the political and socioeconomic realignments that follow disasters. As a historian of the United States, I focus largely (but not exclusively) on U.S. humanitarian and immigration policies in response to these disasters, but the case studies illustrate the necessity and potential for responding to disasters, climate change, and climate-driven migration multilaterally, since our collective future requires collaboration.

Studying the past offers guideposts because humans have reacted to environmental crises in fairly consistent ways over time: natural disasters "have tended to be divisive and sometimes unifying, provoked social and even international conflict, inflame[d] religious turbulence, focus[ed] anger against migrants or minorities, and direct[ed] wrath toward governments for their action or inaction."[7] Humans have also migrated in response to disasters—and will continue to do so—if governments do not minimize risk so people can exercise the right to stay home. Much of the discussion on natural disasters and climate change has focused on sustainability—that is, the "limiting [of] humans' negative impacts on the environment"—and "resilience," the adaptation of life to limit "the environment's negative impacts on human activity."[8] But the discussion must also focus on the accommodation of those unable to adapt to a changing climate. Those who survive sudden-impact extreme weather events or slower-onset environmental degradation are often permanently deprived of home, livelihood, schooling, food, clean water, and sanitation. In order to survive, they must migrate internally within their own countries or across national borders, and the "visceral impacts" of these disasters, such as post-traumatic stress and physical illness, linger long after a disaster has passed.[9] Relocation, then, becomes an adaptive response, but nations often try to block their movement.

Lesser-developed countries—many of which are former colonies whose natural resources and labor were exploited to help colonizers industrialize—now unfairly suffer the most immediate consequences of climate change. The people in these countries are twenty times more likely to be affected by climate-related disasters compared with those living in the industrialized world.[10] Malnutrition, starvation, disease, and human trafficking regularly follow environmental upheaval, making women, children, the elderly, and Indigenous and minority populations especially vulnerable.[11] Children suffer over 80 percent of climate-related deaths.[12] Not surprisingly, we see considerable climate-driven migration within and from—lesser-developed countries.

The scholarship on climate change and climate-driven migration has focused largely on the countries of South and Southeast Asia and sub-Saharan Africa, where we have seen the most dramatic consequences so far, but scholars are now focusing more attention on the countries of the Americas. The Caribbean and Central American countries and territories I discuss in this book—Montserrat, Nicaragua, Honduras, the U.S. Virgin Islands, and Puerto Rico—are responsible for only a tiny fraction of the world's carbon emissions, but climate change has already affected their agricultural production, urban and rural infrastructure, public health, and local and state economies.[13] Geography has made these places particularly vulnerable to sudden-onset disasters like hurricanes, earthquakes, and volcanic eruptions, but so, too, has coloniality, weak governance, indebtedness, short-sighted development policies, and a wide range of non-state actors. Governmental responses often layer disaster upon disaster. Consequently, while most people displaced by disasters around the world have thus far remained in their own countries, Central America has seen among the highest rates of cross-border migration.[14] Drawing on legislative and governmental records, the reports and studies of nongovernmental agencies, and newspapers, photographs, and interviews, I have tried to write a history that examines disasters in the Caribbean and Central America— and the migration that followed—from the perspective not just of the policy makers and aid workers who responded to the crises but also of those who experienced the upheaval on the ground. I examine the policies that exacerbated these so-called natural disasters, shaped domestic and international responses, and changed societies in their wake. The challenges faced by these five countries and territories are in many ways illustrative of the challenges faced by other countries of the Western Hemisphere.

There is general consensus that nations must offer humanitarian relief to populations affected by sudden-onset disasters such as hurricanes and earthquakes and, if necessary, assist with the temporary or permanent relocation and resettlement of those left homeless. There is no consensus, however, on how to respond to slower-onset disasters such as drought and water shortages or policy-driven disasters such as deforestation, acidifying oceans, and toxic waste. These policy-driven disasters, produced by governmental, military, and industry actors, are what Rob Nixon has called a "slow violence": "a violence that occurs gradually and out of sight, a violence of delayed destruction that is dispersed across time and space, an attritional violence that is typically not viewed as violence at all [and is] underrepresented in strategic planning as well as in human memory."[15] These slower-onset disasters have also displaced people and driven them to migrate, as this book discusses, but nations have blocked their movement. Immigration policies worldwide offer few accommodations for those migrants who don't fall neatly into traditional and prescribed legal categories.

As climate change accelerates and environmental disruptions become more violent and frequent, migration across the Americas, and not just from Central America and the Caribbean, will likely increase, and the migrants will follow the historically well-established routes that are the product of geography, cultural and language affinity, and the integration of economic markets. The nations of the Western Hemisphere will have no choice but to respond to this mass migration of people forced to move for economic and cultural survival. The hope is that nations will work collectively to address the challenges that lie ahead.

Are They Refugees?

Those displaced by natural disasters and climate change are often called "climate refugees," but most are not refugees under legal definitions of the term. The term "refugee" has a precise meaning in domestic and international law. The United Nations' *Convention and Protocol Relating to the Status of Refugees* (also called simply the *Refugee Convention*) has defined who or what a refugee is, and that definition has been adopted by 147 countries, including the United States: refugees are individuals unable to return to their country of origin or habitual residence because of a well-founded fear of persecution on account of race, religion, nationality, membership in a particular social group, or political opinion.[16] Nowhere in this definition does one see any mention of natural disasters, the environment, or climate change. In some cases, those displaced by environmental disruptions might be considered refugees *sur place* if they can demonstrate a well-founded fear of persecution because human rights violations erupted in their country after they left.[17] Political upheaval often follows environmental disaster, so many of those displaced might subsequently qualify for legal recognition as refugees and receive assistance with resettlement in another country; but natural disaster, in and of itself, is not a criterion for refugee status. Elsewhere around the world, regional accords such as those drafted by the Organization of African Unity (1969) and the Cartagena Declaration (1984) offer more expansive definitions of refugee status that may offer protections to victims of natural disaster if, for example, their migration was the result of a "serious disturbance to public order."[18]

The United Nations High Commissioner for Refugees (UNHCR) recognizes that the migration caused by environmental disasters and climate change is a type of forced migration but has resisted amending the refugee definition or even using the term "climate refugee" for fear that it will lead to the dilution of protections for the politically persecuted.[19] Instead, the UNHCR and other humanitarian agencies use terms like "climate migrant," "environmental migrant," or "environmentally displaced person."[20] But these terms are not adequate either, because they obscure the political forces that uproot people. Environmental conditions may displace people from their homes, but migrations occur because of the policies of neglect.[21] If nations adopted sustainable development to reduce the impacts of natural disasters and helped their populations recover from—and adapt to—a changing environment (to be "resilient"), fewer people would need to migrate in search of safety and new livelihoods. Unfortunately, the lesser-developed countries from which much migration originates are often highly indebted and do not have the capacity to enact such measures. Migration in this context reveals the political failures of nation-states.

A host of international and regional conventions, protocols, and guidelines acknowledge that people migrate for a wide range of reasons and that they deserve respect and compassion and, at the bare minimum, temporary accommodation. In the past, climate change and climate refugees were absent from these discussions because environmental disruptions such as hurricanes, earthquakes, and drought were long assumed to be part of the "natural" order of life, unlike war and civil unrest, which were considered extraordinary and human-driven and thus avoidable. The expanding awareness that societies are accelerating climate change to life-threatening levels are forcing countries to reevaluate the populations they prioritize for assistance and to adjust their immigration, refugee, and asylum policies accordingly to ensure a just and humane response to those made vulnerable by conditions beyond their control. As the 2005 UN World Summit affirmed, nations have a responsibility to protect (the "R2P principle") in a wide range of contexts.[22]

Some activists have urged the United Nations to revise its refugee convention to recognize a wider range of conditions that create vulnerability, including climate change, to make those affected juridically visible and to offer countries more substantive guidance in shaping immigration policies. As legal scholar Simon Behrman has noted, the term "refugee" has been flexible in the past, in both legal understanding and popular usage, and could be reconceptualized once again.[23] Like the UNHCR, the

International Organization for Migration has viewed such a revision as unnecessary and, instead, has urged countries to draw on existing humanitarian, human rights, and refugee law to offer complementary protection to those displaced by disasters and climate change.[24] Either course of action would be an important acknowledgment that those displaced by climate change have rights and deserve recognition and assistance.[25] It would also help nations define their duties to those who pass through or try to settle in their countries.[26]

The Rights of Migrants and the Duties of Nations

Most people displaced from their homes because of environmental disasters remain in their country of origin, or in the countries that border them, but over time are forced to travel longer distances.[27] Humanitarian agencies have long monitored this internal and cross-border migration. The International Red Cross and Red Crescent reported an exponential increase in disaster-related displacement over the course of just three decades, from 275,000 in the 1970s to 1.2 million in the 1980s to 18 million in the 1990s.[28] More recently, the Internal Displacement Monitoring Center reported that during an eight-year period, 2008 through 2015, over 203 million people were temporarily or permanently displaced from their homes because of flooding, earthquakes, tropical storms, volcanic eruptions, or other natural disasters.[29] While they monitor displacement, humanitarian organizations also underscore the impossibility of forecasting the actual number of people who will be displaced by climate change in the future because so much depends on how nations respond moving forward.[30] What these organizations know for sure is that, at present, most countries offer climate-driven migrants few legal rights, protections, or accommodations.

As early as 1990, the Intergovernmental Panel on Climate Change warned of the likelihood of large-scale migration.[31] Neither the rights of the displaced nor the duties of nations have been addressed in the panel's assessments or in the international climate change agreements that have followed.[32] Human rights advocates have called for clearer guidance on these issues as a matter of social and restorative justice: the developing countries from which migration originates are on the front lines of a crisis that developed nations created through their voracious extraction and consumption of fossil fuels. As one development specialist noted, "Allowing the world's most prolific carbon emitters to ignore the mass displacement of other people would threaten the equality principle of global law."[33]

The UN's *Convention and Protocol Relating to the Status of Refugees* outlines the rights of refugees and the responsibilities of nations to refugees and asylum-seekers, but climate migrants can receive legal recognition as refugees and asylees—and the opportunities for permanent resettlement that come with such status—only if they meet whatever criteria are listed in national statutes. Nations determine who is eligible for refugee status and asylum based on interpretations of their own statutes, but in signing the *Refugee Convention*, countries have theoretically committed themselves to the principle of *non-refoulement*: not returning asylum-seekers to dangerous and repressive conditions. *Non-refoulement* is regarded as the most fundamental of rights.[34]

Those displaced by environmental disaster are not included in the UNHCR's mandate, but António Guterres, UN High Commissioner for Refugees from 2005 to 2015 and later UN secretary-general, has reminded national leaders of the "human imperative" to assist those displaced by climate change. The UNHCR has worked with nations sympathetic to climate change issues such as Norway, which hosted a 2011 conference in Oslo that subsequently led to the drafting of the "Nansen Principles"—ten recommendations for dealing with "disaster-induced displacement."[35] During Guterres's term as high commissioner, the agency also expanded its protection—on an ad hoc basis—for peoples internally displaced within their own countries because of natural disasters.[36] The United Nations has affirmed that its nonbinding *Guiding Principles on Internal Displacement* (1998) covers climate-induced displacement.[37]

There are no guidelines for responding to populations whose countries will disappear altogether, however. The architects of the UN's 1954 *Convention Relating to the Status of Stateless Persons* never imagined that territories might physically disappear because of climate change. Statelessness was defined as the denial of nationality through the operation of law rather than through the physical disappearance of land.[38] The UNHCR protects stateless persons, but it is still unclear at what point citizens of small island states, for example, who are particularly threatened by rising sea levels, might be considered permanently stateless. As McNeill noted, we are in uncharted terrain.

How Has the United States Responded to Climate-Driven Migration?

As one of the wealthier countries in the Western Hemisphere and one of the principal destinations for migrants from all over the Americas, the

United States has profoundly influenced humanitarian and migration policies in the region. The United States has responded to environmental crises on an ad hoc basis and, at this writing, does not have a fully formulated short- or long-term response to climate-driven migration.

Since the establishment of the immigration restriction regime in the late nineteenth century, the bureaucratic tracks for admission to the United States have prioritized those seeking economic advancement, family reunification, or political safe haven. Those who enter the United States as immigrants and temporary workers are generally viewed as economically driven, while refugees and asylum-seekers are said to be politically driven. These distinct tracks fail to recognize that, more often than not, migration is multicausal, produced by a wide and complex range of intersecting drivers that often make it impossible to pinpoint a sole cause for migrating. The environmental drivers of migration are often obscured in the reports of the U.S. immigration bureaucracy. It is only when scholars have conducted oral histories and ethnographies that they have identified the role that environmental disasters have played in uprooting people from their homes. Disasters act as "triggers" and "detonators" of migration, launching a movement of people, but they can also act as "accelerators" of change that is already established and underway; it is in the stories that migrants share that these environmental drivers become visible.[39] People move, temporarily or permanently, because their lives have been in some way upended.

Those displaced by natural disasters are not prioritized for admission to the United States under the "preference system" established by the 1965 Immigration and Nationality Act. Nor are they eligible for refugee status or asylum under the terms of the 1980 Refugee Act unless they are able to demonstrate persecution. There are historical precedents for accommodating victims of disaster, however. U.S. immigration and refugee law once granted protection to populations displaced by environmental disasters. The 1953 Refugee Relief Act, for example, included victims of "natural calamity" in its definition of refugee.[40] Five years later, after a series of volcanic eruptions and earthquakes in 1957–58 devastated the island of Faial, Azores, Congress passed Public Law 85–892, an "Act for the Relief of Certain Distressed Aliens," popularly known as the Azorean Refugees Act, which set aside 1,500 non-quota immigrant visas for "nationals or citizens of Portugal ... who [because of the disaster were] in urgent need of assistance for the essentials of life."[41] The 1965 Immigration Act (Hart-Celler Act) later included in its definition of refugee those persons "uprooted by catastrophic natural calamity as defined by the President who are unable to return to their usual place of abode."[42] By adding "natural calamity" to the refugee definition, Congress hoped to "provide relief in those cases where aliens [were] forced to flee their homes as a result of serious natural disasters, such as earthquakes, volcanic eruptions, tidal waves, and in any similar natural catastrophes."[43]

Between 1953 and 1980, few immigrants and refugees were admitted to the United States as victims of "natural calamity." The Immigration and Naturalization Service (INS) did not include a separate category for disaster victims in its statistical analysis of admissions under the 1953 act, suggesting the numbers were insignificant.[44] Congress extended the Azorean Refugees Act through June 1962, and through this law an estimated 5,000 people entered the United States.[45] Although the 1965 act established an entry track for victims of natural calamity, it also crafted multiple barriers that prevented victims from availing themselves of this opportunity: the law required applicants to present their case to INS officers in one of just eight countries (Austria, Belgium, France, Germany, Greece, Hong Kong, Italy, and Lebanon). If petitioners were successful, they received visas for "conditional entrant status" with no guarantees of permanent residency in the United States unless Congress intervened and provided an adjustment of status. The law also required that the president define "catastrophic natural calamity" and explain how the disaster from which the refugee fled fell within that definition. During the duration of this provision, no president ever defined "catastrophic natural calamity" or declared a disaster to be one.[46]

Despite the small numbers admitted, the "natural calamity" provision was not without controversy. Some legislators said that it was inappropriate to offer permanent resettlement to people who were only temporarily displaced; instead, they argued, the United States should focus on providing material, technical, and other forms of humanitarian assistance to help countries rebuild after disasters. Other legislators took issue with the natural calamity provision because they said it allowed the United States to drain hard-hit countries of their most highly skilled citizens when they were most needed at home.[47] When policy makers introduced an early version of the bill that became the 1980 Refugee Act, they initially retained the natural calamity provision, but Congress subsequently eliminated it from the final law.

At present, Temporary Protected Status (TPS) and Deferred Enforced Departure (DED) are the legal tracks through which the United States accommodates people affected by natural disaster. The 1990 Immigration Act created the statutory provision called TPS: those unable to return to their countries of origin because of an ongoing armed conflict, an environmental disaster, or "extraordinary and temporary conditions" could, under some conditions, remain and work in the United States until the attorney general (after 2003, the secretary of Homeland Security) determined that it was safe to return home.[48] TPS was "the statutory embodiment of safe haven for those aliens who [could] not meet the legal definition of refugee but [were] nonetheless fleeing—or reluctant to return to—potentially dangerous situations."[49] TPS is granted on a six-, twelve-, or eighteen-month basis and can be renewed by the Department of Homeland Security if the qualifying conditions in the homeland persist.

There is one catch, however: one has to be physically present in the United States by a particular date in order to qualify for this status. Congress imposed this requirement to address some legislators' concerns that TPS would lead to a surge in migration from hard-hit areas.[50] Tourists, students, business executives, and those on temporary worker visas are the principal beneficiaries of TPS because they are already in the United States, but undocumented workers can also potentially qualify and receive temporary employment authorization and a stay of removal, unless they are otherwise barred by law (because of misdemeanor or felony records, for example). Individuals with TPS do not qualify for state or federal welfare assistance, but they may apply for "advance parole" that allows them to return to the United States if they must travel abroad for any reason. While they live and work in the United States, TPS recipients are investing in the United States through their labor but also by paying payroll, social security, state, and sales taxes.

Temporary Protected Status is one way, albeit an imperfect one, that the United States has exercised its humanitarian obligations to those displaced for environmental reasons. TPS is based on the understanding that countries in crisis need time to recover; if nationals living abroad abruptly return in large numbers, they can have a destabilizing effect on their country that disrupts that recovery. The U.S. federal government has total discretion in the granting of TPS but expects countries affected by natural disaster to meet two conditions in order for their nationals to qualify: first, the secretary of Homeland Security must determine that there has been a substantial disruption in living conditions as a result of an environmental disaster, making it impossible for a government to accommodate the return of its nationals; and second, the foreign state affected by environmental disaster must officially request TPS for its nationals (this requirement is not imposed on countries that have experienced armed conflict). Once nationals from a country become eligible for TPS, they must apply for this protected status and pay the various fees associated with the adjustment of status that allows them to legally remain in the United States. For some, the high fees make it impossible for them to establish eligibility, and they opt to remain in the United States without legal authorization.[51]

Deferred Enforced Departure also offers a stay of removal from the United States, as well as employment authorization, but DED is most often used as a "placeholder" status after TPS has expired. Through DED, the president exercises discretionary authority to allow certain nationals to remain in the United States if it is in the interest of humanitarian or foreign policy. More often, DED is used when TPS has expired but the White House is evaluating whether to renew TPS or is waiting for Congress to work out the final details of a path to permanent residency and citizenship. By 2005, the president had used DED only six times.[52]

Nicaragua, Honduras, El Salvador, Montserrat, and Haiti are among the twenty-two countries and territories that have qualified for TPS from 1990 to 2020. Salvadorans were the first foreign nationals to receive TPS designation, in response to the brutal war that rocked that country for over a decade.[53] Salvadorans qualified for TPS until 1992, when peace accords were signed and the attorney general announced it was safe for them to return home; but Salvadorans qualified for TPS a second time, in 2001, after three major earthquakes devastated the country. An estimated 212,000 Salvadoran nationals have received TPS, making them the largest group of recipients since the status was created.[54]

The Central American and Caribbean countries and territories on the TPS list have all qualified because of environmental disasters. Montserrat was the first to receive TPS eligibility because of an environmental crisis (which is the subject of chapter 1).[55] When Hurricane Mitch devastated Central America in 1998, Attorney General Janet Reno temporarily suspended the deportation of unauthorized Salvadorans, Guatemalans, Nicaraguans, and Hondurans held in immigrant detention centers, but Nicaraguans and Hondurans were subsequently granted TPS because of the particular devastation in these two countries (see chapter 2).[56] Not all governments that request TPS for their nationals receive it, however, and environmental disaster does not always guarantee TPS designation. The decision-making process is not always transparent or immune from politics. After Hurricane Stan affected Guatemala in October 2005, for example, President Oscar Berger requested that his compatriots in the United States receive TPS, but the George W. Bush administration denied the request; a subsequent request from President Álvaro Colom in 2008 was also denied.[57] No rationale was given.

Critics of TPS have argued that the program runs contrary to Congress's original intent to provide only short-term protection.[58] As early as 1999, a House Judiciary Committee hearing expressed concerns about the program's limited accountability: "The question is not whether TPS should be granted. In many instances, it should be," said committee chairman Lamar Smith of Texas. "The question is whether it is really temporary and to what extent TPS invites fraud. . . . The public has little information about how many aliens have registered for TPS and what happens to them when the TPS designation ends."[59] Officials of the INS testified that the chances of fraud were very low but admitted they did not have the resources to assure that every TPS recipient departed the country once their status expired. The U.S. government did not assist TPS recipients with the costs of repatriation, and without that assistance, and without assurances of jobs and housing in the homeland upon return, it was likely that many opted to remain in the United States, even if it meant living undocumented. Many of those who testified before Congress at the 1999 hearing urged the federal government to work with the UNHCR and other humanitarian agencies to facilitate a safe return and successful economic integration in the home countries: a path the United States has occasionally pursued, as the reader will see in chapter 2.

In some cases, the government has extended TPS for a decade or longer, challenging what some consider to be the original intent of the law. Some international crises are protracted, making it impossible for nationals to return to their country of origin; but despite such prolonged conflicts, TPS recipients have not been offered a pathway to citizenship in the United States. Unless they find some other way to adjust their status and become legal permanent residents, TPS recipients occupy a liminal political space in the United States: unable to exercise the rights of citizenship and with no assurances that they will ever be able to establish permanent roots anywhere. An administration's wide discretion in designating countries for TPS has also raised concerns that domestic political pressure, ideology, and foreign policy concerns might play too great a role in the designation of TPS. Human rights organizations have urged the establishment of clearer standards.[60] In cases of environmental catastrophe, requiring governments to petition for TPS for their nationals also ignores the fact that in "extreme disaster scenarios, the state of origin may be unable to even advocate with other states on behalf of its citizens in distress."[61]

For decades now, scholars and humanitarian advocates in the United States have called for a reevaluation of U.S. immigration and refugee law in order to grant temporary or permanent protection to a wider range of vulnerable populations who are currently excluded from consideration.[62] Advocates have called for the adoption of a more expansive definition of refugee status or for the creation of other complementary humanitarian tracks for admission for those displaced by environmental disasters or climate change.[63] They have also urged policy makers to work with diasporic communities to better assist hard-hit areas by reducing or eliminating the costs associated with sending remittances, for example. As scholars have noted, diasporic communities play an important role in disaster relief and recovery through the transfer of funds, medical supplies, food, and material goods to affected areas.[64] In 2015 alone, migrants worldwide sent $582 billion in remittances to communities back home, including $133 billion from (im)migrants in the United States.[65] Such financial assistance will continue to be critical in the decades to come.

How This Book Is Organized

Chapter 1 examines the particular vulnerabilities and challenges faced by small island states in the Caribbean in the postcolonial world by examining disasters in the British territory of Montserrat. From 1995 to 2010, a series of volcanic eruptions left over half of Montserrat uninhabitable, destroying homes and traditional livelihoods and driving an estimated two-thirds of Montserratians into exile, to neighboring islands such as Antigua-Barbuda, St. Kitts-Nevis, and Guadeloupe but also farther away to the United Kingdom, Canada, and the United States. Paradoxically, a risk assessment conducted a decade earlier had determined that a volcano posed only a 1 percent risk to the population every century.

Although they were residents of a "dependent territory" in the powerful British Commonwealth, the 11,000 Montserratians struggled to receive the humanitarian relief and the post-disaster economic assistance they required to survive and rebuild. The crisis made evident the lack of planning, sustainable development, and culturally sensitive remediation in this territory made vulnerable by its geographic features, but it also illustrated Montserratians' social, economic, and political vulnerability, ruled by dual governments that often worked at cross-purposes. The humanitarian crisis, dubbed the "worst colonial crisis since the Falklands war," contributed to the restructuring of Montserrat's status within the British Commonwealth, and this allowed many exiles to establish permanent residency in the UK (and, theoretically, offered Montserratians a permanent homeland should their island become uninhabitable in the future). Those who sought refuge in the United States became the first foreign nationals to become eligible for TPS for environmental reasons, but as the chapter will show, the status created new forms of vulnerability that demonstrated that TPS is an inadequate response for addressing environmental or climate-driven migration.

Chapter 2 explores how U.S. foreign policy interests have shaped responses to disasters and disaster-driven migration by examining Hurricane Mitch's impacts on Honduras and Nicaragua in 1998. At this writing, Mitch is the deadliest hurricane in Central American history, and Honduras and Nicaragua were the two countries that experienced the highest casualties and greatest physical devastation. Hurricane Mitch struck a comparatively small geographic area in the Western Hemisphere, but its impacts were felt across the region. In the first three months after the storm, the United States spent $300 million on rescue and relief operations, and billions more over the next few years, not just out of humanitarian concern but because aid and reconstruction served foreign policy objectives: humanitarian aid would protect the fragile peace secured in the early 1990s, after decades of wars in the region. The United States' involvement in Central America had deep roots, and this complicated history shaped U.S. responses to Hurricane Mitch and to those displaced by this disaster. In the two decades that followed, neoliberal economic development policies designed to assist Central America's recovery produced slower-onset disasters that have also threatened lives and livelihoods, displacing even more Central Americans from their homes. Back-to-back political and environmental disasters in these so-called post-conflict societies have made full recovery and resilience difficult and have reinvigorated migration.

In both chapters 1 and 2, we see the limitations of Temporary Protected Status as a response to cross-border migration. TPS is the principal legal instrument through which the United States has accommodated those unable to return home because of natural disasters. The Montserratians who were granted TPS numbered just a few hundred people, and their chances for reintegration back home were limited because of the protracted environmental crisis; nevertheless, after eight years, the George W. Bush administration revoked their protected status, prompting a public outcry and a legislative battle that is discussed in chapter 1. Similarly, the U.S. foreign assistance package for Central America after Hurricane Mitch involved offering TPS to tens of thousands of Hondurans and Nicaraguans who were already living in the United States at the time of the disaster, many of them in the country without authorization and facing deportation. Though they were a significantly larger population than the Montserratians, the federal government repeatedly extended TPS for the Hondurans and Nicaraguans. The Central American TPS holders have been allowed to live, study, and work in the United States for over two decades. Despite their long presence in—and ties to—the United States, many have not found a pathway to citizenship, however. In this context, TPS has created a liminal population that, for a quarter century, has been both part of the nation and yet legally on its margins. By 2019, Nicaraguans and Hondurans, together with Salvadorans who received TPS in 2001, accounted for 80 percent of all TPS recipients in the United States.

The Montserratian, Nicaraguan, and Honduran case studies reveal the difficulties of assessing who merits TPS protection, when it is safe for TPS recipients to return to their homelands, and how politicized that process of evaluation can become. While TPS offers recipients the chance to remain in the United States for a period of time and to send much-needed remittances to hard-hit areas to assist recovery, it can also contribute to insecurity, making it difficult for TPS holders to plan for the future or to belong anywhere. These case studies raise an important question: If conditions in the home country prevent TPS recipients from returning, at what point should they be offered a pathway to permanent residency and citizenship?

Chapter 3 looks at a very different case study: the back-to-back hurricanes of 2017 that devastated the American territories of the U.S. Virgin Islands and Puerto Rico. As citizens of one of the richest and most powerful nations of the hemisphere, Virgin Islanders and Puerto Ricans did not require TPS or any other immigration status to receive safe haven. Theirs was a form of internal migration: they could migrate freely to, and remain indefinitely in, the continental United States without passports, visas, or the payment of exorbitant administrative fees. In the wake of these hurricanes, tens of thousands made use of these privileges and migrated to Florida, Texas, New York, Pennsylvania, and other states, following routes established by other Virgin Islanders and Puerto Ricans since the nineteenth century.

As U.S. citizens, Virgin Islanders and Puerto Ricans should have qualified for the same economic assistance as other Americans who experience natural disaster, but they competed with other hard-hit areas of the United States for diminishing economic resources. By the late summer of 2017, financial coffers for disaster relief and recovery were already close to depletion because of 692 federally declared disasters and emergency declarations. Congress subsidized the Federal Emergency Management Agency's Disaster Relief Fund based on forecasts of disaster activity, but the scale of the 2017 disasters was "unprecedented," FEMA said, and required Congress to pass additional emergency appropriations bills so that FEMA and other agencies and departments could address these various environmental crises. FEMA later reported that the environmental crises of 2017 had "tested the nation's ability to respond to and recover from multiple concurrent disasters."⁶⁶ Policy makers hoped—perhaps futilely in an era of accelerating climate change—that there would not be any more years like 2017.

The islands' political status as "unincorporated territories" also proved to be a liability, barring them not only from full economic assistance from the federal government because of statutory caps placed on assistance to U.S. territories but also from certain types of international assistance. The physical and economic devastation that these U.S. territories experienced because of Hurricanes Irma and Maria cannot be attributed solely to nature's wrath. The hurricanes may have destroyed homes, factories, farms, businesses, and public utilities, but the policies of Washington and Wall Street in the decades leading to—and in the wake of—2017 exacerbated the destruction, made recovery difficult, and forced tens of thousands to consider migration their only option. Hurricanes Irma and Maria revealed, yet again, that the American citizens of U.S. territories occupy a liminal and unequal political status in the American empire that makes economic growth difficult and their recovery from natural disaster less of a priority for federal policy makers. Addressing these gaps in assistance requires reassessing entrenched political and economic relationships.

A final chapter offers some reflections on humanitarian and immigration policies in response to environmental disasters and climate change and on the prospects of international cooperation. Natural disasters may be impossible to prevent, but much of the devastation that occurs in their wake is policy-driven and can be mitigated with sound planning and preparation. The climate forecasts for the next fifty years are daunting, making climate change one of the most important challenges in world affairs today. Every nation is—and will continue to be—affected by accelerating climate change and will have its own climate migrants. It is hard to imagine that nations confronting their own environmental challenges and internal migration will adopt generous immigration policies to accommodate those temporarily or permanently displaced from their homes, but climate change is a global challenge that requires nations to work together and share any burden collectively. The best response to climate-driven migration is not ad hoc immigration policies, which are only bandages on a much larger problem. Nations must address the reasons people migrate in the first place—and that involves difficult conversations about the economic policies that have accelerated climate change, indebted nations, driven more people into poverty, and made it impossible for people to exercise the right to stay home. Only then can we begin to help populations adapt to the challenges of today and tomorrow.

CHAPTER 1

What a Volcano Revealed about the Vulnerability of Small Island States

> This volcano likes to give us surprises. The longer the quiet went on, the more certain we were that we had seen the end. But apparently it wasn't done yet.
>
> —Gill Norton, Montserrat Volcano Observatory

> The top must-sees are not your predictable, nice-and-easy kind of places. They may puzzle you, make you do double-takes, even unsettle.
>
> —Montserrat tourism website

On July 18, 1995, a long-dormant volcano in Soufrière Hills on the Caribbean island of Montserrat showed the first worrisome signs of steam venting and ashfall. A month later, on August 21, an eruption occurred on Chance's Peak—the first major volcanic eruption on the island in almost four centuries—forcing the relocation of 5,000 people, most of them to the northern part of the island. Within a few weeks, the evacuees were allowed to return to their homes, but by late November, the sighting of pyroclastic flow forced a second evacuation. Montserratians returned to their homes again in early 1996, but by April, a third evacuation was necessary.

Over the next two decades, volcanic activity—including eruptions in 1997, 2000, 2001, 2003, 2004, 2006, 2009, and 2010—caused widespread devastation. With a total land area of only 103 square kilometers (or roughly 40 square miles), the phreatic surges and flows damaged or destroyed three-quarters of the island's infrastructure, damaging water supplies, electrical lines, and telephone service, as well as homes, businesses, medical clinics, schools, government buildings, the airport, and even the state archives that documented centuries of Montserrat's history and culture.[1] The 1997 eruptions left the capital city of Plymouth uninhabitable and required the relocation of the seat of government, first to Salem and eventually to the northwestern town of Brades. The environmental crisis brought to a halt most administrative, commercial, industrial, and seaport activities; devastated the agricultural, fishery, and tourism sectors of the economy; and at one point or another led to the temporary or permanent relocation of 90 percent of the island's roughly 10,000 residents. Many of the evacuees were relocated more than once and forced to live in spartan conditions in emergency shelters, with little or no access to electricity, refrigeration for food, sanitary disposal of waste, or privacy.[2] Respiratory and other health problems, along with the psychological stress of coping with the seemingly interminable disaster, exacerbated Montserratians' difficulties. According to one U.S. State Department report, the population was "at some risk of contracting lung disease from inhaling airborne particles contained in the volcanic ash" that covered the island.[3] By 1998, three-quarters of Montserratians had left the island for the United Kingdom, the United States, Canada, and various countries in the Caribbean. Most of the evacuees would never again return to their homes.

Environmental and Political Hazards in the Postcolonial World

Montserrat is one of the smaller Leeward Islands of the Lesser Antilles. In 1493 Christopher Columbus named the island after the Benedictine abbey on the outskirts of Barcelona, Santa María de Montserrat. The island became a British colony in 1632, populated largely by the Crown's disposable populations—prisoners, widows, and orphans—as well as by enslaved Africans and Anglo-Irish Catholic settlers from Barbados, Virginia, and St. Kitts.[4] The island developed a plantation economy, growing sugar primarily, but also tobacco, indigo, cotton, and limes. Sugar production and the plantation system declined after the abolition of slavery in 1834. Agricultural production slowly shifted to smaller landholdings that were passed on from generation to generation, but many of the island's villages retained the names of the colonial plantation estates. After the Second World War, tourism, light manufacturing, and construction emerged as key sectors of the economy. North American and European tourist-expatriates, seeking retirement homes, sparked and financed a construction boom on the island that lasted into the late 1980s.[5] Tourists called Montserrat the "Emerald Isle of the Caribbean" for its Irish heritage, Irish-named landmarks, and verdant landscape (it is the only area outside of Ireland that celebrates St. Patrick's Day as a national holiday, and visitors' passports are stamped with a shamrock).[6] The island became a favorite destination for musicians Paul McCartney; Elton John; Stevie Wonder; Sting; Jimmy Buffett; Michael Jackson; Earth, Wind, and Fire; and the Rolling Stones, who recorded their music at George Martin's Associated Independent Recording Studio in Plymouth.[7] Many of these musicians would later perform at a benefit concert at Royal Albert Hall in London to raise funds for Montserratians' emergency relief.

Except for a few brief periods when the French occupied the island, Montserrat was a "Crown colony" and later, in 1967, a "dependent territory" of the United Kingdom, with an appointed governor that mediated between elected local legislators and the UK Foreign Office. The appointed governor oversaw external affairs, defense, internal security (including police), public service, offshore finance, and the judiciary; until 2010, when Montserrat adopted a new constitution, the governor had discretionary authority to appoint officials and to intervene in Montserrat's internal affairs when deemed necessary.[8] In the midst of the humanitarian crisis of 1995–2004, Parliament passed the British Overseas Territory Act of 2002, which reclassified Montserrat as one of Britain's fourteen self-governing "Overseas Territories" in the Commonwealth;[9] and 4 years later—and almost 400 years after its colonization—Parliament granted Montserratians British citizenship. Other accommodations would follow after 2010. Despite these political changes, the United Nations has included Montserrat on its list of non-self-governing territories, and each year a UN General Assembly resolution reaffirms Montserratian's right to self-governance.

Although the events of 1995 caught the local population—and the Montserratian and British governments—by surprise, the volcanic eruption was not unusual for the Caribbean. Montserrat is one of a long chain of islands of volcanic origin, from Saba in the north to Grenada in the south, and its geographic position on fault lines makes it vulnerable to volcanological and earthquake activity. The French settlers of the Caribbean called the volcanic areas *soufrières*, and twenty of these Caribbean

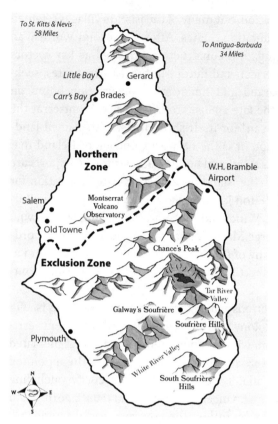

Montserrat. (Map by John Wyatt Greenlee)

volcanoes remain active today. During 1897–1900, 1933–37, and 1966–67 a series of "subvolcanic earthquakes" shook Montserrat but produced little evidence of active magma.[10] More deadly volcanic eruptions occurred in Martinique and St. Vincent in 1902: in Martinique, the debris from a lava dome on Montagne Pelée swept the town of St. Pierre and claimed the lives of over 29,000; and in St. Vincent, the falling ash from the soufrière killed 1,680.[11] Volcanic activity in Guadeloupe in 1976 forced the temporary evacuation of 72,000 people, and in 1979 another volcanic eruption in St. Vincent displaced 10,000 people.[12] Ships are barred from a maritime exclusion zone between St. Vincent and Grenada because an underground volcano called "Kick 'em Jenny" frequently demonstrates seismic activity.

Volcanoes may be a common feature of some parts of the Caribbean, but Montserrat had not witnessed an eruption in centuries. Indeed, a 1992 travel feature in the *Toronto Star* told potential tourists that they could safely visit a clearing at the edge of the "steaming, sulphorous center of an inactive volcano" to "peer into the crater from a natural platform." There was no fee, no railing, no sign telling visitors about its geological history, said the article. "None of the things that change a natural phenomenon into a tourist attraction. You can walk along a 20-minute trail right up to the bubbling mud and hot springs."[13]

According to scientists, most volcanoes reach a peak relatively early in their activity cycles and then gradually decline, but Montserrat's Soufrière Hills showed no signs of slowing down. The 1995 eruption was followed by protracted periods of dome growth and collapse, and the scientists who monitored the area warned government officials that at any moment pyroclastic flows of superheated gases, rocks, and boulders, with a potential ground speed of 200 mph, could incinerate everything in their path.[14]

Despite its history of inactivity, the UK had commissioned a risk assessment of Soufrière Hills in 1987, which determined it posed a "1 percent risk" to the population every century. Perhaps because of the exceptionally low risk, the plans to minimize potential damage were never enacted; indeed, the assessment was later reported lost by three departments.[15] Not surprisingly, this British territory was not prepared to deal with the consequences of the 1995 volcanic eruption, especially since Montserrat was just barely recovering from Hurricane Hugo, another major disaster just a few years earlier. Hugo, a Category 4 storm when it reached Montserrat in September 1989, had disrupted agricultural production, destroyed or heavily damaged over 90 percent of the buildings on the island, and left a quarter of the population homeless. But on the eve of the 1995 eruption, the government had reported that Montserratians were on the verge of full economic recovery, in large part due to financial assistance from the United Kingdom, which had helped them restore their normally good standards of housing, education, and health care services.[16] The volcanic eruption of 1995 undermined this recovery. As journalist Polly Pattullo wrote, "The builders were still painting the new library in Plymouth, the new government headquarters had yet to hold a legislative council meeting and the improved hospital had yet to receive a patient" when the volcano erupted.[17] Although volcano-tectonic activity near Montserrat was detected as early as April 1989, "none of the detected precursors was a clear, unambiguous indicator of an imminent eruption."[18] Consequently, there was no contingency plan, and ad hoc arrangements were put in place "reactively as the eruption progressed."[19]

The dual political hierarchies—Montserrat and the United Kingdom—made emergency responses, as well as the short- and long-term investments in hazard mitigation, difficult to enact. While the governor had the constitutional authority to take control of governmental matters in times of crisis, various political actors in the Montserratian government and in the UK Department for International Development (DFID) argued over who had jurisdictional authority over the decisions that had to be taken. The consequence was "delays, omissions and shortcomings," especially during the early days and months of the crisis. Facing the uncertainty of continued volcanic eruptions, as some scientists warned, the government of Montserrat tried to maintain a semblance of normality "to avoid jeopardizing the long-term viability of the island community."[20] The British government, in turn, adopted a "wait and see" approach during the first eighteen months of the crisis and provided only "very basic health, shelter, and social assistance" to those who had to be evacuated from their homes.[21]

The island's small size, the population's concentration on the flanks of the active volcano, and the slow and unpredictable dome growth were among the many challenges of emergency management and planning. A team of over a hundred scientific, technical, and clerical staff from Montserrat, Trinidad, Puerto Rico, the United Kingdom, France, and the United States arrived to monitor the volcanic activity, assess and map the hazards and risks, and communicate them to the general public.[22] The Montserrat Volcano Observatory, founded soon thereafter, helped the government of Montserrat develop an alert system to disseminate information through radio, newspapers and magazines, public meetings, school visits, and even poster boards. Through this public education campaign, volcanic terminology became well understood by local communities and widely used in public debates.[23]

Despite this public information campaign, it was often difficult to convince local communities that they had to evacuate, especially if residents were elderly or infirm. Montserratian police established checkpoints around the continually expanding high-risk "exclusion zone" to prevent locals from entering and occasionally sent staff door-to-door to warn area residents of impending danger. Despite these precautions and admonitions, many people found ways to enter or remain in the exclusion zone, in part because they distrusted or misunderstood the scientific information or believed the volcano would give them sufficient audible or visual warning of an impending dome collapse so they could escape in time. Others returned to or remained in the exclusion zone because they felt they had little choice: they had land, pets, and livestock to tend to. They viewed a permanent relocation as out of the question because of their generational ties to the land and the limited opportunities for housing and fertile land elsewhere on the island.[24] Montserratians' greatest fear, wrote geographer Tracey Skelton, was the loss of their land, which had been "handed down through past generations [and had] the potential to be places for the future generations of their own families.... It was a symbol of freedom."[25]

The decision to return to their property in the exclusion zone proved costly. In May 1997, scientists once again documented dome growth and collapse and warned that one of these would cause pyroclastic flows down the northern flanks of Soufrière Hills. Montserratians in this area were urged to evacuate, but for some, the fact that the local airport remained open sent a contradictory message that conditions were not as dire as the scientists had forecasted.[26] Others either didn't have radios or stopped listening to the twice-daily reports from the Montserrat Volcano Observatory. On June 25, 1997, the predicted pyroclastic flows destroyed 200 homes in seven villages in the southern part of the island. More than a thousand people were evacuated but nineteen people died, trapped in their homes, cars, and fields, as molten lava and ash enveloped them. The casualties ranged in age from three months to seventy-three years of age. Many more suffered serious burns and were airlifted to hospitals around the Caribbean. These were the first deaths directly attributable to the eruption, but in the days that followed, officials recorded other deaths that Montserratians believed were indirectly related.[27]

By early July, the pyroclastic flows encroached on the capital city of Plymouth. Explosions continued into October. One of the largest dome collapses occurred on December 26, 1997, when the southwestern flank of Soufrière Hills gave way and moved down the White River valley as a debris avalanche. A large volcanic ash cloud and a pyroclastic flow of hot gas and volcanic debris quickly followed. Ash fell as far away as the island of Guadaloupe, and substantial debris entered the sea, forming a small tsunami that threatened neighboring islands.[28]

The 1997 eruptions forced DFID, which meets the "reasonable assistance needs" of Britain's overseas territories, to take more proactive measures to assist with evacuation.[29] Prior to this date, the British government had instituted only an informal voluntary evacuation plan, in part to appease the government of Montserrat, which feared that relocation to other countries would lead to Montserrat's permanent depopulation. Under the terms of the voluntary plan, Montserratians who made their own way to the UK were allowed to stay temporarily, for up to two years if they had a sponsor, with work authorization and access to public education, but received no financial assistance. The severity of the 1997 eruption now required a more comprehensive response: the UK government now offered Montserratians transportation to the neighboring islands of Antigua-Barbuda and Guadeloupe, where they were offered housing, financial assistance, and medical care; the amount of financial assistance varied depending on destination but averaged £2,400 for every adult and £600 for every child under eighteen, to be paid in installments over six months. The Royal Navy warship HMS *Liverpool* anchored off the west coast of the island to facilitate the evacuation. Those on the island who wished to settle in the UK were promised the "right of abode" and work authorization, as well as assistance grants to help with the costs associated with transportation, resettlement, and education for their children. They were also now entitled to social welfare benefits, if needed.

DFID's actions were not without controversy. Those who chose to remain on the island had comparatively less support than those who chose to leave. One common complaint was the lack of safe housing for those forced to relocate to the northern part of the island. Those unable to move in with family and friends either became squatters in churches and schools or lived in the flimsy tents British authorities set up in parks, which often blew away during stormy weather. In these tent communities, Montserratians cooked over open stoves. The only sanitation consisted of open-pit latrines that regularly overflowed.[30] British authorities eventually shipped in "container-style metal boxes with windows and doors that let in little light and almost no air or breeze."[31] As many as two dozen people were assigned to live in these hot metal structures, their living quarters separated by wooden partitions. Local authorities called these dwellings "socially, culturally and structurally unacceptable"—worse than the stalls where the British kept their cattle, said some—and yet many evacuees were forced to live in such structures for years because of the lack of affordable housing on the island. While they remained in these conditions, they had to deal with other natural disasters: five hurricanes swept through the area in 1995 and 1996 alone. Montserrat's chief minister, David Brandt, blamed the growing death toll on "the failure to make reasonably adequate housing available two years after the crisis [began]," which contributed to deteriorating health conditions and forced many people to defy orders and return to their homes in the exclusion zone.[32]

Compounding Montserratians' anxieties was the announcement that the insurance companies that operated on the island planned to withdraw altogether or issue notices of nonrenewal of coverage, undermining Montserratians' chances of financial compensation for loss of property. Call-in programs on Radio Montserrat became an outlet for the population's frustration and anger over the bureaucratic inefficiencies, the lack of basic social services, and the apparent devaluation of life.[33] Montserratian politicians were not exempt from criticism: the widespread protests over his handling of the crisis led to the resignation of Chief Minister Bertrand Osborne, Brandt's predecessor.

British government officials had their own complaints about Montserratian legislators, the obstacles they created, and the inefficiencies in the delivery of aid. Claire Short, the newly appointed secretary of state for international development, complained about Brandt, in particular, and his ever-growing "shopping list" of demands, when Montserrat already commanded so much of Britain's foreign aid budget. One interview on August 24, 1997, just weeks after nineteen people were killed, was especially controversial: "My department's budget is designed to help the poorest people on the Earth, and I have to be very responsible about how it is spent. It would be weak politics if I said, 'They are making a noise and a row. Oh dear, give them more money.' They say 10,000, double, treble and then think of another number. It will be golden elephants next. They have to stop this game. It is bad governance. It is hysterical scaremongering, which is whipping people up."[34] Short eventually apologized for the comment, but the interview made her deeply unpopular on the island.

Anti-British resentment grew further when a British delegation in October surveyed the damage for only one day and granted the chief minister only a thirty-minute meeting.[35] Other incidents elicited complaints. The twenty-member Caribbean Community, or CARICOM, offered to build thirty houses on the northern end of the island, in "Caribbean style" to accommodate "the extended and complex household formations found in the region"; but when CARICOM asked British authorities for an extension of electric cables from an adjoining housing project, the British bureaucrats stalled for over a month while they studied the request, despite the urgent need for housing. The country of Nevis also tried to assist Montserrat's recovery by offering evacuees free land to build new homes; when Nevisians asked the British government to fund the construction materials, British officials refused.[36] "The real issue here," wrote Brandt in a letter to *The Guardian*, "is the continuously obstructive manner of U.K. government officials. . . . In the meantime, thousands of Montserratians are suffering appalling conditions and cannot understand the obstruction of essential aid."[37] The first permanent housing units (fifty properties) weren't parceled out until November 1997, two years and four months after the crisis began, and only to those able to pay the rent.[38]

These episodes reinforced Montserratians' perceptions that their country, like other dependent territories, simply wasn't a priority for the

British government. British civil servants thousands of miles away in London either misunderstood—or were unsympathetic to—the scale of the emergency. Prime Minister Tony Blair reassured Montserratians that Her Majesty's Government remained "fully committed to [the] viability of the future of Montserrat"; but because financial assistance to those who left the island far exceeded that given to those who remained, rumors circulated that the British government planned to use the environmental crisis as a pretext to either abandon the island it considered unproductive or depopulate it in order to lessen its financial obligations to its sustainable development.

Responding to public pressure, in 1997 and 1998 the UK House of Commons Select Committee on International Development investigated the government's handling of the "worst colonial crisis since the Falklands War."[39] The report from the committee was highly critical of the foreign secretary and DFID for their handling of the Montserrat volcano disaster, which they said led to "unnecessary tensions and inefficiencies" and demonstrated a "lack of political will."[40] Since DFID had failed to heed earlier recommendations, the committee proposed that another government office in Whitehall assume oversight of the overseas territories so that future relief and rescue operations could be handled more expeditiously.

DFID conducted its own internal study of the Montserrat crisis. Its two-volume report (1999) arrived at a very different conclusion, the authors judging the work on the island "a success in comparison with many other recent natural disasters elsewhere in the developing world": "There were only 19 confirmed fatalities directly attributable to the eruption and hardly any measurable increase in communicable disease and physical ill health.... Throughout the emergency, involving four major evacuations at little notice, everyone has had a roof over their head, no one has gone hungry and there have been no reported cases of child malnutrition, and social order has been maintained."[41] The report also listed as DFID successes the enhancement of scientific monitoring of volcanic activity on the island and the sustainable development plans in place for 1998–2001.

The report did admit that reconstruction was behind schedule and that Montserrat's economy was "virtually nonexistent," which meant that the majority of the on-island population was at least partially dependent on social assistance until livelihoods and commerce could be restored. The total capital loss, including real estate, was unofficially estimated at £1 billion, much of it unrecoverable.[42] As the DFID report stated, "The considerable achievement of the people of Montserrat is to have coped with the continuing volcanic menace that was wholly outside their experience, and then to adapt to the loss of homes and livelihoods and the disruption to their community."[43] Montserratians were less forgiving: a 1999 inquest into the nineteen deaths concluded that in nine of the cases, the failure of the British and Montserrat governments to provide healthy accommodations for displaced farmers contributed directly to their deaths.[44] Against orders, those killed had returned to look after their livestock or to get the fruit and vegetables that were in short supply in their refugee settlements and were later trapped.

Over the next few years, DFID supplemented its £59 million aid package to the island (1995–98) with an additional £75 million (1998–2001), which eventually made Montserrat the second-largest recipient of Britain's international aid during this period.[45] DFID authorized a "Sustainable Development Plan" and "Country Policy Plan" to outline the objectives, timetables, and budgets for the recovery of every sector of Montserratian society, from education, health, and tourism to infrastructure and construction. Short- and long-term planning was difficult given the uncertainties of future population size, but government officials hoped economic investments in development would lure evacuees back to Montserrat. The government of Antigua-Barbuda also received financial compensation for temporarily (or, in some cases, permanently) accommodating 2,500 Montserratians during these years, but Anguilla and St. Kitts-Nevis, which accommodated fewer evacuees, did not.[46]

Montserrat's volcanic crisis made evident the island's "ambiguous and fragile political situation as a British colony" and contributed to a subsequent restructuring of the United Kingdom's relationship to its dependent territories (or, as some called them, the "remnants of empire").[47] In March 1999, stressing the historic ties between the UK and its "overseas territories," the British government released a "white paper" examining the idea of a "partnership" that more clearly defined the obligations and responsibilities of the various parties in the postcolonial world.[48] While some considered the proposal a "reinscription of colonialism . . . constructed around a political rhetoric of partnership," the changes did lead to tangible benefits for Montserratians. Prior to 2002, those with "British Overseas Territory Citizenship," though British nationals, continued to be subject to immigration controls: they did not have the automatic right to live and work in the UK, nor were they considered UK nationals by the European Union.[49] But beginning in 2002, residents of thirteen British Overseas Territories, including Montserrat, finally received British citizenship and with it the right of abode. Many observers found it surprising that a vigorous independence movement did not emerge on the island in the wake of the volcanic eruptions, but this political "partnership" appeared to be a positive choice for Montserratians and, as Tracey Skelton noted, "complicat[ed] the assumption that all colonial territories *ipso facto* seek independence."[50]

The status renegotiations, sustainable development plans, and financial aid packages did not end Montserratians' concerns for their long-term security. In 2000, the lava dome at Soufrière Hills produced ash clouds that shot up as high as 30,000 feet and blanketed even parts of the island's designated "safe zone."[51] The following year, the lava dome once again spewed "volcanic mudflow, ash, small stones, and hundreds of tons of sulphur dioxide." The July 2003 eruption was the largest eruption since the volcano emerged from dormancy in 1995.[52] In March 2004, yet another eruption "sent a massive cloud of ash into the air" and "pyroclastic flows down the eastern flank of the Soufrière Hills," settling volcanic ash—up to four inches in some areas—and spreading volcanic grit on nearby Caribbean islands.[53] All of these eruptions once again caused major damage to the island's infrastructure, as well as to agriculture and fisheries: after the 2003 eruption, for example, 95 percent of the island's crops were destroyed, including fruit trees and seedlings cultivated in nurseries; livestock were affected with respiratory ailments; and dead fish and birds washed up along the coastline. Even the Montserrat oriole, the national bird, was absent from the island and was considered close to extinction, along with other bird and bat species. The rationing of water and the distribution of masks once again became necessary.[54]

By 2005, two-thirds of the island was uninhabitable, and roughly 5,000 of the island's residents had resettled abroad.[55] Those who remained in Montserrat faced loss of income or high levels of unemployment due to the impacts on agriculture and tourism; with reconstruction occurring at a slow pace, only a privileged few found jobs in construction.[56] A de facto capital was rebuilt: first in the village of Salem, where businesses, banks, and government offices reopened—some of them in the villas owned by absentee expatriates—and subsequently in Brades, when Salem also had to be evacuated. "I don't think there's any country in living memory that has lost two-thirds of its population and two-thirds of its land space, its only hospital, airport, seaport and thousands of homes," said Chief Minister Brandt. "We're [rebuilding] from scratch."[57]

The volcano has remained comparatively quiet since its last significant eruption in 2010, but scientists have recorded more than 100 earthquakes around the volcano since 2007. Montserrat's volcano remains one of the most closely studied in the world. The Montserrat Volcano Observatory's

Hazard Level System, accessible online, now routinely alerts the population on the soufrière's potential threats.[58]

Accommodating Environmental Migrants

Montserratians have a long tradition of migrating to countries in the Caribbean basin in search of work. Between 1845 and 1930, thousands migrated temporarily or permanently to Trinidad, St. Vincent, Guyana, Panama, the Dominican Republic, Cuba, Aruba, and Curaçao to work in the sugarcane fields and the oil refineries and in the building of the Panama Canal. In his 1973 study, anthropologist Stuart B. Philpott argued that Montserrat was a "migration-oriented society" and "highly dependent on those of its members living and working abroad." Close to 5,000 Montserratians traveled to the UK between 1946 and 1964, with migration increasing most dramatically after 1954. The majority earned their living in factory work, transportation, or service occupations, and their remittances bettered the life chances of family members left behind. An additional 2,600 settled in the United States and the British and Dutch West Indies.[59] According to the historian Violet Showers Johnson, Montserratians and other Afro-Caribbean peoples grew into a visible presence in the city of Boston, then the headquarters of the United Fruit Company (now Chiquita Brands), until the 1952 McCarran-Walter Act created obstacles to that migration.[60] Even though migration led to a decrease in Montserrat's population, in the immediate postwar decades the island recorded its largest population size, reaching a high of 14,233 in 1955.[61]

Despite this migratory tradition, the post-1995 exodus was unparalleled because of the large numbers of Montserratians exiting in a very short period. According to scholar Gertrude Shotte, this migration changed the social, cultural, and political landscape of the territory.[62] By 1999, roughly two-thirds of Montserrat's population had left the island, driven out because of conditions brought on by the volcanic eruptions: high unemployment, the lack of affordable housing, and the skyrocketing costs of food and services. When families moved abroad, many of Montserrat's institutions and businesses shut down, contributing to more unemployment and a higher cost of living. Prior to August 1995, there were 2,672 students enrolled in Montserrat's schools, for example, but by 1998 only 620 remained; the number of teachers also declined, from 200 in 1995 to 54 in 1998. "How can we live here with no banks, no insurance, no shops, no homes?" asked one resident. "The island's only dentist left at the weekend, one of the two remaining pharmacies said it was going and Britain [says it will close] down the one remaining hospital."[63]

An estimated 3,500 people, or roughly half of the exiles, settled in the UK by 1999, especially in North and East London, where other Montserratians had relocated in the past, and in Nottingham and Leicester. The other half dispersed around the Caribbean, especially to Antigua-Barbuda and St. Kitts-Nevis, where bilateral agreements allowed Montserratians to live and work, as well as to Canada and the United States.[64] Those who settled in the UK prior to the 2002 restructuring of Montserrat's political relationship encountered the harsh practical realities that, despite their territory's historic ties to the British Commonwealth, they were technically "visitors," and this made their long-term status and economic prospects in the UK precarious. City governments at first settled Montserratian families in temporary accommodations like nursing homes and "bed and breakfasts" before moving them to houses and flats in "colonies" like Top Valley and The Meadows in the city of Nottingham so they could offer one another support. Volunteers raised funds to provide the new residents with kitchenware, electrical appliances, furniture, and winter clothing.

A year after arriving in the UK, many Montserratians had still not found jobs, because their work experience and qualifications went unrecognized or they lacked the educational credentials required. This placed high-paying skilled jobs out of reach.[65] Even when they found employment, their long-term future in the UK remained uncertain. In a conversation with a British Foreign Service officer, one man summed up Montserratians' frustrations: "We have been British for over 400 years. We fought the First and Second World Wars for Britain. Men lost their lives for Britain. We are supposed to belong to you and here you tell us that we have to ask for the right to stay in this country. I find that hard. I think we should know our identity, whether we are British or not. Tell us now."[66] In Canada, by comparison, which was also part of the British Commonwealth, officials from Citizenship and Immigration Canada were instructed to "treat favorably [Montserratians'] requests for extensions of status, to facilitate employment authorizations . . . to facilitate opportunities for seasonal agricultural workers, and to accelerate the processing of applications for permanent residence."[67]

Smaller countries in the Caribbean, with less-developed economies, were perhaps even more generous. Antigua, for example, had long allowed Montserratian workers to seek employment on the island without work permits, but it now granted Montserratian evacuees permission to stay and work long-term and to receive education and health services, even though the arrival of 2,500 Montserratians in a short period of time had strained its social security systems, despite British financial assistance. Montserrat also received contributions of food, prefabricated housing, medical supplies, and volunteer labor from its fellow CARICOM nations.[68]

Those who traveled to the United States and its Caribbean territories arrived with immigrant, tourist, and humanitarian parole visas, and they settled in communities established by previous waves of Montserratians and other Caribbean immigrants. In 1990, the Montserratian immigrant population in the United States stood at roughly 3,500 people, and most were concentrated in the Boston and New York metropolitan areas. Some 1,016 Montserratians legally immigrated to the United States between 1990 and 1999 (426 between 1995 and 1999), most through the "family preference" category of immigration law that facilitates family reunification. Only 2 Montserratians were granted refugee status during this period (for reasons unknown). Of the over 1,000 Montserratian immigrants who arrived during the 1990s, 470 settled in the New York metropolitan area.[69] Regardless of where they settled, Montserratian communities raised money and sent goods back home to assist in the island's recovery.[70]

Montserratians who arrived in the United States (or its territories) on tourist, student, or other temporary visas often overstayed their visas and joined the large populations of unauthorized workers in these cities who worked primarily in the service economies. In 1997, Montserratians became the first foreign nationals to qualify for Temporary Protected Status for reasons of environmental catastrophe, and those in the United States with or without authorization qualified. According to government records, the Department of Justice authorized TPS because "there exist[ed] extraordinary *and temporary* [emphasis mine] conditions in Montserrat that prevent[ed] aliens ... from returning to the island in safety." Although the Department of Justice estimated that as many as 1,000 Montserratians might be eligible for TPS, the exact numbers are hard to come by. The total of those who came to avail themselves of this legal protection may have been much smaller, either because the government overestimated their numbers or, more likely, undocumented Montserratians feared coming forward, making themselves visible to the government and subsequently vulnerable to removal. The Justice Department initially granted TPS for only one year, from August 28, 1997, to August 27, 1998, because the Clinton administration believed that the damage on the island, though devastating and substantial, would cause only a temporary disruption of living conditions there.[71]

A year later, when the environmental disaster had not abated and the Montserrat government seemed unable to accommodate returning nationals, Attorney General Janet Reno extended TPS a second time, through August 27, 1999. The George W. Bush administration later extended protected status four additional times, noting the continued threat from further volcanic eruptions, the housing shortage, and the serious health risks from volcanic ash. State Department memoranda always explained the reasons for the extension. The 2000 memorandum, for example, reported that "the island has remained in a state of crisis . . . and [since March 2000] the volcano [has] turned deadly again . . . [and] recent reports from the center that monitors the volcano's activity indicate that another such event may occur very soon."[72] Each extension of TPS required the original recipients to reregister for this temporary protection, even if they were trying to secure permanent residency through some other legal means.

The new Department of Homeland Security (DHS), created in 2002, and which now assumed jurisdiction over TPS, eventually conceded that "the volcanic eruptions [could] no longer be considered temporary in nature" and concluded that TPS was no longer a viable response. The DHS cited as evidence one study by the Scientific Advisory Committee on Montserrat Volcanic Activity, which estimated that "there [was] only a 3.2% chance that this period of volcanic activity [would] stop within the next six months. There [was] a 50% probability that the volcanic activity [would] last another 14–15 years, and a 5% chance that the volcanic activity [would] continue for over 180 years." Consequently, "because the volcanic eruptions [were] unlikely to cease in the foreseeable future, [Montserratians could] no longer be considered 'temporary' as required by Congress when it enacted the TPS statute," said the announcement. On August 27, 2004, the DHS announced a phasing out of Montserratians' TPS, granting them an additional six months of work authorization "to provide for an orderly transition" that would lead up to the official termination date of February 27, 2005, when Montserratians would be ordered removed from the United States unless they secured some other type of visa that allowed them to remain. The DHS urged Montserratians to exercise their new rights to British citizenship and settle in the UK.[73]

The announcement that TPS would be terminated elicited considerable shock and even outrage. Chief Minister John Osborne, who succeeded Brandt, warned that their small island did not yet have the housing or jobs to accommodate those who returned and urged the British government, which was also caught by surprise, to use its leverage to persuade Washington to reverse the decision. "We feel that the American government is kind of hard hearted given our circumstances to be imposing a situation like this on the Montserrat government at this time," said Osborne.⁷⁴ Editorials and op-eds across the United States also criticized the decision. "In the tortured world of bureaucratic reasoning," said the *Patriot Ledger* of Quincy, Massachusetts, "Montserrat is still an inhospitable place. But because it will remain so for the foreseeable future, the 'temporary protected status' that applied to Montserrat refugees will not be extended. Got it? The situation that drove them to the U.S. is no longer temporary. So they can no longer claim protected status as refugees—even though the government acknowledges Montserrat still poses serious health risks because of air pollution from ash." If the families were from Cuba rather than Montserrat, wrote the editorialists, "no questions would be asked. This is just one more example of a refugee policy that is subjective and unequal."⁷⁵

An editorial in the *Washington Post* noted, "Why exactly DHS has taken this absurd and cruel step is a bit of mystery. The Montserrat refugees are hardly a drain on American society. And the crisis that led them to flee their homes was not one of their own making. While DHS officials argue that the permanent solution is for them to go to Great Britain, which rules the island, why is it reasonable or humane to uproot them at all?" In articles and op-eds, the Montserratians were often called refugees because they were seeking safe haven, but they did not meet the criterion for *legal* refugee status. The *Washington Post* recommended that there be some process by which people who have been in the United States for years under the TPS program become permanent residents: "It is wrong for this country—having promised people haven—to renege on that promise because the problem that drove them here turns out to be more severe and more tractable than policymakers imagined."⁷⁶

Some Montserratians found a way to adjust their status to legal permanent resident (often through marriage or through an employment visa): in 1999, for example, an estimated 1,000 Montserratians held TPS, but by 2003 U.S. Citizenship and Immigration Services data showed that only 325 held this status, suggesting most had either found a pathway to permanent residency, left the United States, or chosen not to renew their TPS.⁷⁷ A year later, only 292 Montserratians held TPS, but it was unlikely they would be able to adjust their status before the February deadline. Those ordered removed from the United States would have to take their American-born children with them, but "not because it [was] safe to go home again. . . . It [was] not going to be safe anytime soon."⁷⁸ Because Montserrat's airport was still not fully functional, returning to their homeland required flying into Antigua and then taking a ferry, seaplane, or helicopter to the island.

"I'm not going to leave," said one Montserratian in Boston. "We've got to stay and fight the battle because I don't have anywhere to go."[79]

In the months leading to the February deadline, political allies looked for other solutions. Thirty-six members of the Congressional Black Caucus, which during the 1990s had also taken up the cause of Haitian and other Black asylum-seekers, urged the president to grant permanent residency to the 292 Montserratians facing removal. "If deported," their letter said, "these individuals will face many challenges including being homeless, unemployed, and in jeopardy to their physical and mental health sentence as received. They pose no threat to our country, unless being an educator, a nurse, a subway station attendant or technician are being deemed subversive activities."[80] In the Montserratian case, the Congressional Black Caucus never directly accused the Bush administration of racism, as its members had when criticizing Haitian asylum policy, but other critics did wonder if the administration would have been more accommodating of the 292 individuals had they been white or fleeing a communist adversary like Cuba.

Having anticipated the possibility of removal, in 2001 Congressman Major Owens of New York introduced H.R. 1726, a bill to provide for the adjustment of Montserratians' immigration status from TPS to legal permanent resident. Despite the very small number of individuals who stood to profit from this adjustment, the bill died in committee. Owens reintroduced the bill in 2003 and then again in 2005, but in the post-9/11 anti-immigrant climate, virtually all immigration bills were difficult to pass.[81]

Congressman Stephen F. Lynch of Massachusetts was also among those who came out in support of the bill, arguing that forcing Montserratians to return to their island endangered their lives and undermined an economy not yet recovered from the ongoing disaster: "To expel the Montserratians now . . . [would be] an affront to our long-standing reputation as a leader in refugee protection. Many of these men and women have been living and working in the United States for a decade, and have become important members of our communities and our churches. And we'd be forcing their U.S. citizen children to leave their schools and friends and the only home they've ever known. It's just wrong to punish legal immigrants who are working hard, paying taxes and playing by the rules."[82] Lynch's comments underscored all the points that might convince even the wariest of constituents: Montserratians were law-abiding and gainfully employed and believed in family, faith, and education. This argument, like Owens's, failed to convince a majority of legislators to pass the bill. Legally, the Montserratians were visitors, not "immigrants," and legislators did not feel compelled to take up their cause.

In the Senate, Charles Schumer of New York also introduced a bill to authorize an adjustment of status—the Montserrat Immigration Fairness Act of 2004—and when the bill failed to pass, he reintroduced it the following year.[83] As the TPS termination date approached, Schumer, along with Senators Edward Kennedy and John Kerry, wrote President Bush asking that the Montserratians be granted Deferred Enforced Departure for an additional twelve to eighteen months so that they could at least continue to work, save money, and make arrangements for new lives back home or in the United Kingdom: "Whatever choices the Montserratian refugees are forced to make, they will need to have some means to support themselves for at least six months or more after they leave the United States, and this deferred period would enable them to prepare for their departure, prevent some economic and social hardships for themselves and their families, and provide them with an opportunity to seek a safe haven and jobs outside the United States."[84]

The Bush administration chose not to pursue these recommendations. The termination date arrived and the Justice Department ordered Montserratians to leave the United States. Donna Christen-Christensen, the delegate from the U.S. Virgin Islands in the U.S. House of Representatives, told news reporters, "We have done everything we could, including sponsor legislation in both the House and U.S. Senate to get the White House and the Department of Homeland Security to change their minds . . . but our pleas have sadly fallen on deaf ears."[85]

It is unclear how many Montserratians returned to Montserrat or settled permanently in the UK after TPS was rescinded. Those who chose to remain in the United States without protected status joined the millions of other migrants who lived and worked undocumented in the service economy. In doing so, they assumed great risks. Under the terms of the 1996 Illegal Immigration Reform and Immigrant Responsibility Act, anyone apprehended by U.S. immigration authorities faced incarceration and deportation and was subsequently barred from reentering the United States for a period of up to ten years. How many chose to take that risk is impossible to determine from immigration records.

Reevaluating Sustainable Development and "Resilience" in a Small Island State

In a May 1999 meeting with Montserratians in London, Chief Minister David Brandt urged those eager to return home not to do so unless they had guaranteed housing and employment. "We do not have anywhere to put you," he told his compatriots and warned them that the government could provide social services only to those who had remained on the island throughout the crisis.[86]

Despite these admonitions, Montserratians did take advantage of the British government's "Assisted Return Passage Scheme," established in June 1999 to help Monserratians return home. Those who went encountered an island much changed. Fewer than 5,000 people remained on the island, many of them elderly: their children and grandchildren had abandoned Montserrat in pursuit of opportunities they could not find in their homeland. Two-thirds of the island—the "Special Vulnerable Area" or "exclusion zone" around the volcano—was charred, uninhabitable, and abandoned; and three areas along the coasts, the largest extending four kilometers, were designated "maritime exclusion zones" where shipping could not enter.[87] In the old capital city of Plymouth, streets and buildings lay buried under meters of hardened mud. Many of the old village communities were gone, and the population was now concentrated in "housing estates" in the north.[88] Prior to the 1995 eruption, most Montserratians owned their own land and homes, but many were now renters, which exacerbated their feelings of dislocation.[89] More worrisome yet, the volcano continued to expel hot gas and debris, compromising their health and undermining whatever economic recovery their politicians pursued. Twenty years after the first eruption, many Montserratians continued to live in emergency shelters, and roughly half of Montserrat's children remained in poverty in an island that had once enjoyed one of the highest standards of living in the Caribbean.[90]

British private interests profited from Montserrat's recovery and reconstruction plans. The *Montreal Gazette* reported that "the multimillion-dollar contract to build new homes went to a British contractor; the thousands of dollars Britain [spent] daily on [the] jet-catamaran ferry service that links Montserrat to the world [went] to a company with British investors; and a large share of Montserrat's aid [was] used to finance the salaries, travel expenses and recreation leaves of British officials based here." But Montserratian subcontractors also reaped some financial benefits, as did the "small army of guest workers" who arrived from Jamaica, Guyana, St. Vincent, Dominica, Haiti, and the Dominican Republic to take advantage of the construction boom that eventually followed.[91] By 2002, roughly half of the people living in Montserrat were migrants from these neighboring islands, and "without them the country could not function," reported the International Organization for Migration.[92] But Montserratians blamed these migrants for depressing wages, driving their young out

Sign warning of danger from volcano in the Montserrat exclusion zone. (Photo by t. m. urban/Shutterstock.com)

of their homeland, and increasing crime.[93] Some were also suspected of simply using their residency in Montserrat as a stepping stone for acquiring British citizenship and the right of abode in the UK, since a year after acquiring Montserratian citizenship they became eligible for British citizenship.[94] Not surprisingly, the newcomers often complained that they were socially marginalized.

Professional occupations and more skilled positions went unfilled because many with specialized training had abandoned the island. "[Montserratians] are here against indescribable odds," said one development officer to journalist Polly Pattullo. "I detest the word resilience—we're not resilient, we're stressed and burdened. You have to have an incredible will to live here, because sometimes we just want to run away. . . . What we've achieved is unbelievable because there are so few of us."[95] Perhaps the greatest ongoing challenge was establishing a sense of community in the wake of so much dislocation and change. "So many people have had a complete change in their lifestyle, which was basically a good one," said one medical officer. "They owned their own houses, had reasonable space, gardens to tend. They had their friends around them, and extended family. All that's been totally disrupted." Post-traumatic stress levels were high during the 1995–2004 period, especially among the children, and Montserrat's government contracted psychologists to come to the island to assist the population.[96] "You can rebuild physical damage," Chief Minister Brandt said. "But the emotional scars from the separation from your family, after having lost everything in the twinkling of an eye, the feeling when you, as an independent person, suddenly become dependent on the government for everything—when you live through that kind of indignity, it must leave a scar."[97]

In the aftermath of the eruptions, Montserratians relied on the food, clothing, medical supplies, and remittances their relatives abroad sent home, even if these remittances were too small to merit documentation by the World Bank and other international economic institutions. International agencies like the Red Cross and the United Nations' Children's Fund also aided recovery by providing support to migrants and longtime residents, assisting with debris removal, assessing vulnerability to natural hazards, and implementing mitigation projects.[98] Montserrat received financial assistance from the European Union, the Caribbean Development Bank, CARICOM, the United Nations Development Program, and the Organization of Eastern Caribbean States, but the British government offered the most substantial assistance—£350 million by 2012.[99]

Montserrat's long-term recovery required a reevaluation of what constituted sustainable development in an island increasingly threatened by hurricanes, earthquakes, and volcanic eruptions, as well as by population loss. As an "overseas territory," Montserrat's political relationship with—and economic dependency on—the United Kingdom always complicated this reevaluation. Many felt that, during the 1995–2004 period, DFID and the government of Montserrat could have done more to involve the population in the decision-making processes. The result "might not [have been] the kind of economic development envisaged through neoliberal economic discourses," wrote geographer Tracey Skelton. "However, it might [have served as an] alternative development that allow[ed] Montserratians to remain on their island, rebuild what matters to them and move towards sustainable practices, however small-scale that might be."[100]

After 2004, Montserratian officials more proactively solicited the input of its 4,882 citizens in designing and implementing its *Montserrat Sustainable Development Plan 2008–2020* (*SDP*) and *Physical Development Plan for North Montserrat, 2012–2022* (*PDP*). Development should secure for Montserratians "self-sufficiency," "economic independence," and population growth, said the *SDP*, but this required more housing, jobs, and commerce, concentrated in a smaller area of the island. In order to accomplish these goals, the *SDP* proposed shifting "the drive behind economic activity" from the government to the private sector, which, it said, was more capable of creating jobs and opportunities both for residents on the island and for members of the diaspora who longed to return home. "Independence in [Montserrat's] economic affairs" required economic, social, and institutional frameworks that "facilitate[d] private sector growth, good governance, effective management of the risks associated with natural hazards and climate change, and the sustainable use of [Montserratian] natural resources."[101]

Complementing the *SDP* was the *Physical Development Plan*, which not only offered a blueprint for disaster mitigation and environmental management but also addressed good governance and economic growth. According to the *PDP*, economic development had to focus on the expansion of agriculture, tourism, renewable energy, mining, and manufacturing. Residential, commercial, and industrial development in the north of the island had to be "green" and sustainable, and the *PDP* proposed building a new capital and economic center at Little Bay and a new maritime port at Carr's Bay. To meet national food production targets, the *PDP* proposed implementing agricultural techniques that would reduce the amount of land required and the incidence of insect pest damage to vulnerable crops. The plan also recognized that Montserratians had to protect the island's "ecology, natural resources, and beautiful vistas as well as its built heritage." These were "important not just for maintaining a high quality of life for the island's residents . . . [but to] attract visitors and returning Montserratians."[102] With these plans in place, Montserratian officials believed they could increase the population to 9,007 by 2022 while minimizing their vulnerability to natural disasters and climate change. To guarantee jobs for returning Montserratians, the island ended its open-door employment policy for nationals of the CARICOM countries. The Office of the Premier also convened a "consultation" of the Montserratian diaspora in the town of Little Bay to acquaint returning nationals (and all others interested in investing in Montserrat) with the various measures the government was taking; the consultation later published a twenty-seven-page "handbook" with practical information and advice pertinent to resettlement.[103]

To jump-start tourism and commerce and place the territory on the road to self-sufficiency, the British government invested millions in a new regional airport at Gerald, in the north-central part of the island, primarily for small planes that could navigate the shorter runway. Critics of the plan argued that the two-hour ferry service from Antigua was a more realistic

Plymouth, the old capital of Montserrat, buried under mud and lava flows from Soufrière Hills volcano. (Photo by James Davis Photography /Shutterstock.com)

option for an island with such a small population, but British officials insisted the expenditure was warranted.[104] Ironically, the volcano became critical to Montserrat's development plans. Tourism companies capitalized on "disaster tourism," catering to those who wanted to see the "mysterious Montserrat" with its still-active Soufrière Hills, the Montserrat Volcano Observatory, and the ruins of Plymouth, which became the "star attraction." Tourism brochures, articles, and websites call the former capital city the "Pompeii-in-the-tropics," a "ghost town," and "the only volcanic-buried town in the Americas," where "houses, shops, and churches have been reclaimed by vegetation but front doors are open and entering is an experience that is both eerie and poignant." "Mid-90's Caribbean fashions fill bedroom wardrobes and living room furniture stands intact, if a little dusty," said one tourism article. "Kitchen cupboards are still filled with jars of sauces and condiments that might have been opened that evening and used for dinner, if someone hadn't knocked on the door and told the inhabitants to run and get as far away as possible."[105] Over the next decade, tourism slowly returned to the island. By 2019, 21,000 tourists had visited

Montserrat, the first time since the 1995 eruption that the number had exceeded 20,000.[106]

The volcano also became important for geothermal energy, quarrying, and sand mining. Companies began mining in the exclusion zone, hoping to make some profit off the otherwise unpopulated and unproductive areas of the island. The DFID's investments in Montserrat's energy grid allowed the island to transition away from diesel-powered generators to geothermal's renewable energy. This was an especially important model, because, as one journalist noted, "all of the islands of the Lesser Antilles have similar geological settings and therefore geothermal potential."[107] Together with tourism, these new economic sectors became sources of jobs and income in the post-eruption economy, supplementing the traditional and more modest exports of electronic components, plastic bags, apparel, hot peppers, limes, and live plants. It is too early to tell if these development plans will allow Montserrat to achieve self-sustaining economic and population growth. By 2020, Montserrat recorded a population of 4,992, up from its all-time low of 4,835 in 2005, but still far smaller than the population the *SDP* and *PDP* had hoped to attract.[108]

In the decades to come, Montserrat's security will continue to be threatened not only by volcanic and seismic activity but also by the intensifying storms that come with rising ocean temperatures. The forecasts for small island states like Montserrat are sobering: coastal erosion and saltwater intrusion from sea level rise; flooding and wind damage from more violent and unpredictable storms; drought from higher temperatures; and agricultural loss due to increased infestation. Even the volcano itself might be affected by higher sea levels since geologists have noted a historic correlation between retreating glaciers and increased volcanic activity.[109]

Like other small island states, Montserrat is preparing as best it can for these forecasts. The Montserrat Volcano Observatory continues to monitor volcanic activity. The state-owned radio station provides regular scientific updates on the ashfalls and volcanic gases, and an island-wide siren system, tested daily, warns the population whenever they must take precautions.[110] Working with the Organization of Eastern Caribbean States, Montserrat has adopted a "Climate Change Policy and Action Plan" for disaster risk mitigation.[111] The government has relocated critical and vulnerable infrastructure and implemented a new building code. Farmers are experimenting with new crops and cultivation methods that have proven to be more effective in resisting high temperatures and drought. Through "renewable energy and the new energy policy, the new building code, [the] conservation of the Centre Hills biodiversity, and agricultural extension,"

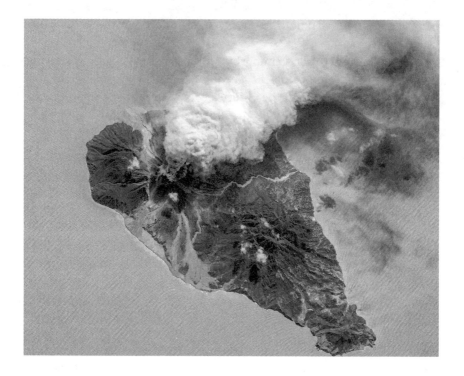

Ash and steam plume, Soufrière Hills volcano, October 11, 2009. (Photo by Jeffrey Williams, NASA)

the territory hopes to stay one step ahead of forecasted environmental crises.[112] And if adaptation efforts fail and homes and livelihoods cannot be preserved or restored quickly, Montserrat's Disaster Management Coordinating Agency has contingency plans in place for the evacuation of the population.

What Next?

Residents of small island states like Montserrat have lived with the reality of storms and other natural disasters for generations. These are part of their history, but in an era of accelerated climate change, Montserrat offers a cautionary tale on the need to prepare for a very different environment even when risk assessments suggest safety. Nations can no longer respond to crises ad hoc. Plans must be implemented, not drafted and then shelved.

Confronting climate change, residents of small island states will likely experience environmental disruptions more regularly, making it even more challenging for populations to fully recover. In 2017, Montserrat avoided the devastation caused by three back-to-back Category 5 hurricanes that destroyed other islands in the northern Caribbean, including Barbuda, which had given Montserratians refuge in the 1990s, but residents know that many more powerful storms are in their future. Doomsayers argue that these adaptation and mitigation efforts are simply buying Montserratians (and other small island states) time and that eventually the entire population will have to be relocated. In 2021, the International Organization for Migration began assessing the capacity of Montserratians and other Caribbean populations to relocate and the resources needed to ensure a safe and orderly migration.[113] If the need for permanent relocation were to come, Montserratians' British citizenship—and their very small numbers—may allow them to resettle in the United Kingdom or around the British Commonwealth. Neighboring islands facing the same risks and challenges, but with much larger populations and sovereign status, will not have that option. They will need other countries in the region to help accommodate their populations, as they helped Montserratians during their time of crisis.

As in many environmental crises, Montserratians displaced by the volcano at first migrated internally, concentrating in the north, but over half the population was eventually forced to migrate elsewhere. Migrants from other parts of the Caribbean subsequently arrived in search of low-wage jobs created by the post-disaster construction-based economy, but their numbers never replenished the population. These workers have been economically and socially marginalized and denied a voice in government and in decision-making. The experiences of Montserratians (and the migrant workers who have settled in their homeland) point to the few protections in place for those forced to migrate for reasons of environmental catastrophe and raise important questions about the rights of displaced persons and the duties of nations. The UK was a logical destination for Monserratians to resettle given their historic, political, economic, cultural, and linguistic ties, and yet Montserratians' accommodation was not automatic. The right to resettle in the UK had to be fought for, recognized, and granted. Those who settled in the United States, though significantly smaller in number, found their opportunities for permanent residency or extended visitor status blocked. Was the forced return of Montserratians a violation of the principle of *non-refoulement*? Bush administration officials argued "no" because Montserratians had another viable option for resettlement, namely, the United Kingdom, where they held citizenship; but for the 292 Montserratians who spent eight years working, paying taxes, and establishing community in the United States, moving to the UK was yet another upheaval. Removal was unnecessary and even callous.

Environmental disasters do not always lead to out-migration. Two devastating hurricanes in 2008 did not trigger a large-scale exodus from Haiti, for example. The tipping point came two years later, on January 12, 2010, when on the heels of two hurricanes an earthquake measuring 7.0 on the Richter scale killed an estimated 250,000 people, left 3 million people homeless, and produced $14 billion in damages.[114] Only then did Haitians migrate in sizable numbers. Like the Montserratians of 1997, Haitians who were in the United States at the time of the earthquake qualified for the protections of TPS. Their numbers were significantly larger, however, because Haitians had been migrating to the United States for decades. Of the estimated 110,000 unauthorized Haitian immigrants in the United States, 46,000 received TPS because the conditions in Haiti in the aftermath of the earthquake made their return impossible. But like the Montserratians, by 2019 they, too, were on the cusp of losing their protected status because an administration seemed frustrated by the country's failure to achieve basic levels of recovery.[115] The Montserratian and Haitian cases raised the important question: At what point should individuals on TPS be allowed to make a case for legal permanent residency?

The volcanic eruptions in Montserrat revealed the geographic, political, and economic vulnerabilities of one small island state in the Caribbean. As a result of this environmental and humanitarian crisis, Montserrat lost over half of its population and also much of its habitable territory. Governmental and industry policy makers are now focusing on sustainable and climate-resilient development, which is critical to this now smaller territory's future viability; but should these programs fail to protect against the hurricanes, earthquakes, and volcanic eruptions that are endemic to the region, contingency plans are also necessary to assist with the evacuation of its population and its temporary or permanent relocation elsewhere.

CHAPTER 2

Disaster Relief as Foreign Policy

When Poverty, Conflict, and Catastrophe Collide

> I have seen earthquakes, droughts, two wars, cyclones and tidal waves, but this is undoubtedly the worst thing that I have ever seen.
>
> —Cardinal Miguel Obando y Bravo, Roman Catholic archbishop of Managua, Nicaragua

> This is the worst disaster to befall Honduras in a hundred years. This has been a harder blow to us than all the 100 military coups we've suffered in our history put together, harder than all the 36 civil wars we've gone through put together.
>
> —William Handal Raudales, vice president of Honduras

The Central American isthmus is one of the most environmentally challenged areas of the Americas. This thin stretch of land is vulnerable to the hurricanes and tropical storms that form in both the Pacific Ocean and the Caribbean Sea. Over a twenty-year period, from 1996 to 2016, twenty-eight major hurricanes[1] and twenty tropical storms directly hit one or more Central American countries.[2] During this same period, many more hurricanes and storms passed within sufficient proximity to cause surges, flooding, and other physical damage.

Volcanic eruptions, earthquakes, and windstorms are also endemic to the isthmus. The countries of the region form part of the Central American Volcanic Arc: a chain of 100 volcanoes roughly 900 miles long, most of them concentrated in Guatemala, Nicaragua, El Salvador, and Costa Rica, some of which erupt on a regular basis. Five of Nicaragua's twenty-one volcanoes, for example, erupted forty-eight times between 2001 and 2017.[3] One of the deadliest earthquakes of the late twentieth century occurred in December 1972, when an earthquake measuring 6.3 on the Richter scale struck Nicaragua. Although the epicenter of the quake was located twenty-three miles northeast of Managua, it destroyed much of the capital city, killing 10,000 people and leaving a quarter of a million people homeless.

Drought is yet another challenging feature of this area. El Salvador, Guatemala, Honduras, and Nicaragua form part of the Central American "Dry Corridor," a geographic area stretching from southern Mexico to Panama that is particularly vulnerable to El Niño–Southern Oscillation events. It is not uncommon to see intense rain and flooding in one sector of the isthmus and drought in an adjacent area. Whether by flooding or drought, these natural disasters affect agricultural production and cause food insecurity for the over 1 million workers who rely on seasonal and subsistence farming.[4]

Central American countries have demonstrated varying capacities in recovering from natural disasters. The isthmus is home to some of the poorest and most indebted countries in Latin America. Among the more progressive is Costa Rica, where despite a low GDP, government policies have expanded access to social services and contributed to life expectancy while reducing income inequality and ecological impact.[5] Honduras, Guatemala, and El Salvador, on the other hand, are the least developed, in part because they are still grappling with the social, economic, and political fallout of the wars of the 1970s–90s. The impacts of natural disasters are felt more keenly in post-conflict societies and in countries with weak governance and already high levels of inequality. The rural poor are disproportionately affected, as this chapter makes evident.

For decades now, men and women have fled the environmental disruptions and food insecurity in Central America by migrating internally within their own countries or across international borders to adjacent countries or to *el norte* to find work and to send remittances back to their families.[6] As one anthropologist noted, migration has become an adaptive response to upheaval, "a path to development, rather than a symptom of its failure."[7] The escalating levels of violence in the first decades of the twenty-first century, driven by government corruption, paramilitaries, gangs, and drug trafficking but also by private industry's crackdown on environmental activism, have only increased the number of people who have sought safety and better economic opportunities in industrialized countries like Mexico, the United States, and Canada. Central American migration is therefore propelled by multiple causes—environmental, economic, and political. The immigration bureaucracies of developed nations like the United States, however, categorize these migrants as economic migrants rather than as political asylum-seekers or climate refugees, which, in a world of highly secured borders, makes them ineligible for admission. The environmental drivers of the migration are lost on the policy makers who block their entrance.

This chapter examines one environmental catastrophe—Hurricane Mitch in 1998—and its impact on migration from Nicaragua and Honduras. According to the National Climatic Data Center, Hurricane Mitch was the deadliest Atlantic hurricane since 1780, causing "human and property damage on a scale never before experienced in the recorded history of this hemisphere."[8] Central Americans called Mitch "El Monstro": the hurricane killed 11,000 people, left close to 3 million people homeless, and caused billions of dollars in damages in Nicaragua, Honduras, Guatemala, and El Salvador.[9] Of these four countries, Nicaragua and Honduras were the hardest hit.[10] Hurricane Mitch also prompted one of "the largest relief operation[s] ever directed at any natural disaster in [U.S.] history."[11] In prior decades, from 1964 to 1998, the United States had spent a total of $297 million addressing various disasters in Central America; Hurricane Mitch alone required over $1 billion for rescue and relief operations and recovery programs.[12] In the decades following Mitch, the United States has spent billions more for relief efforts in the region, helping countries recover from earthquakes, volcanic eruptions, drought, and more Category 4 and 5 storms.

An examination of Hurricane Mitch offers an accounting of the economic, social, and political fallout that countries and regions experience when struck by catastrophic storms and other natural disasters. Hurricane Mitch may have struck a comparatively small geographic area in the Western Hemisphere, but its impacts were felt across the region. Hurricanes like Mitch may be impossible to prevent in an era of accelerating climate change, but much of the devastation that occurs in the wake of natural disasters is due to negligence or policy failures and can be mitigated through political will. Mitch offers lessons, should nations wish to heed them, on the importance of development programs that enlist the participation of civil society, allow citizens to rebuild quickly in the wake

Honduras and Nicaragua. (Map by John Wyatt Greenlee)

of disasters, and do not create or exacerbate vulnerability. Mitch can also instruct governments on the need for regional burden-sharing and the importance of implementing migration policies that temporarily or permanently absorb a share of those most affected by disasters.

Disasters Are Never Just about the Forces of Nature

Central America's rainy season begins in May and can last as long as December; tropical depressions and hurricanes are common during these months.[13] On October 24, Mitch became the thirteenth hurricane of the 1998 season. By the afternoon of October 26, Mitch had reached the northern coast of Nicaragua and Honduras as a Category 5 hurricane with a surface wind velocity of 290 kmh (180 mph) causing storm surges of over eighteen feet, torrential rainfall, flooding, and mudslides. Over the next few days, the hurricane advanced southward and inland. By the time the storm reached the Honduran capital of Tegucigalpa in the geographic center of the isthmus (October 30), Mitch had been downgraded to a tropical storm, but during that five-day period, the storm had flooded the coastal plain and had caused landslides in the mountain areas.[14] On October 31, a weakened Mitch appeared to be moving toward the northern

Pacific Ocean but strengthened in wind velocity once again and changed course, this time moving across El Salvador and Guatemala. It eventually crossed the Isthmus of Tehuantepec in Mexico, entered the Gulf of Mexico, crossed south Florida, and finally weakened in the North Atlantic.

The disaster made evident the poor city planning, inferior building construction, and unsustainable agricultural, forestry, and ranching practices in many parts of Central America. It also made obvious the deep-rooted inequality there, since it was the rural and urban poor—especially women, children, the elderly, and Indigenous and Afro-descendant peoples—who were disproportionately killed or uprooted from their homes, lands, and livelihoods. The communities of the poor, built with inferior materials on riverbanks, lakeshores, hillsides, and mountain slopes, were especially vulnerable to the flooding and landslides. Deforestation, one of the many consequences of the decades-long exploitation of natural resources in the name of economic development and modernization, had created unstable slopes that now prevented a natural control of the rain currents and augmented the scale of the disaster.[15]

In Nicaragua, the largest country in Central America, 450,000 people were left homeless, but over 860,000—or 19 percent of the 1998 population—were affected by the hurricane in some way. The government declared nine of the country's sixteen provinces disaster areas, but the western departments of León and Chinandega, where the most vulnerable sectors of the population lived, suffered the most severe damages. Several towns were swept away when the crater lake atop the Casita volcano collapsed and an avalanche of mud and debris came crashing down, covering sixty square miles below. "We heard a boom from the mountain, and immediately after, an avalanche of mud carried everything away," Posoltega mayor Felicitas Zeledón said. "People tried to escape, but they were swept away along with trees and animals."[16]

The first rescue missions to reach the area found the remains of men, women, and children, their heads and arms "stuck out of the mud as if pleading for help." Houses were submerged in the debris. Over half of the 3,800 Nicaraguans killed by the avalanche were children. For months afterward, over a thousand people remained missing and were presumed dead.[17]

Across the country, the catalog of destruction was heartbreaking. The bridge of land that separated Lake Managua (also known as Xolotlán) from Lake Nicaragua (Cocilbolca) twenty miles to the south disappeared under water, creating one large lake that swallowed the communities on its banks. The town of Tipitapa, on that bridge of land, was subsumed under seven feet of water. The storm destroyed tens of thousands of private and public buildings, including houses, schools, and medical clinics. The country's hydroelectric, thermoelectric, and geothermal energy plants sustained severe damage, as did the state-owned Nicaraguan Telecommunications Company.[18] Although emergency humanitarian aid arrived from all over the world, aid workers found it difficult to reach needy victims because the storm had destroyed or damaged bridges and rendered roads impassable.

To makes matters even more dire, the flooding and mudslides unearthed civil war–era land mines that injured eight people and killed four (by March 1999), including sixteen-year-old Bernardo Ocampo González, who died after a floating mine exploded as he bathed in a river.[19] Over the previous decade, the Nicaraguan government, working with the Organization of American States, had removed 100,000 land mines along the Nicaraguan-Honduran border, where both the Sandinista military and U.S.-backed Contra rebels had planted them during the 1980s to deter troop movement; but as many as 75,000 land mines were yet to be removed. The flooding and mudslides now altered their locations.[20]

Storm survivors were afflicted with a variety of physical ailments. The Nicaraguan health ministry reported that many now suffered from acute respiratory infections, diarrhea, swamp fever, conjunctivitis, foot fungus, and skin diseases. Vector-transmitted diseases such as cholera, malaria, and dengue expanded in the months following the hurricane, straining the capacity of health care facilities. Survivors also suffered from post-traumatic depression, mourning the loss of children, spouses and parents, homes and livelihoods. "When a plane passes in the night, the children start crying and screaming," said one of the elected leaders of a camp for displaced persons. "They think it's another mudslide." Psychologist Josefina Murillo Vargas remarked that the most heartrending cases she encountered in her relief work in Nicaragua were the children and adolescents left orphaned: "The children cried out for their parents even though they knew they had died."[21] With the loss of parents, children suddenly became the heads of their nuclear families. "I talked to one boy [thirteen years of age] who lost his parents and five brothers and sisters out of eight," said Roberto Aguilar, the director of mental health for Nicaragua's health ministry. "He ticked the names off to me, one by one, but he didn't cry or even change expression. I asked him, 'Aren't you sad?' And he said, 'Now that I'm the man of the house, I can't cry.'"[22]

At the time of the hurricane, Nicaragua was regarded by economists as the second-poorest nation in the hemisphere, with close to half of the population unemployed or underemployed. As a result of the storm, 70 percent of the working population lost their livelihoods. In the northern region of the country, farmers suffered heavy losses in export crops such as bananas, sesame, sugarcane, and peanuts. In the Pacific region, the banana and sugar crops were affected, while in the north-central region, basic grain crops, tubers, and vegetable crops rotted or washed away. Some 13,000 hectares of rice could not be harvested that year. The following year's harvest of corn, rice, and beans was also affected, since planting generally occurred in November and December, but fields and equipment remained underwater for months.

Other sectors of the Nicaraguan economy also suffered. Dairy, egg, and meat production declined over the next year because of the loss of livestock. The shrimping and fishing industries were hard hit when shrimp farms were damaged or destroyed and fishermen lost their boats for trawling. Broken sewer lines and septic tanks polluted area waters, making whitefish, another staple product, unsafe to eat and unsafe for export.[23] Only the coffee crop seemed comparatively unscathed, but the condition of roads, bridges, and ports made delivery of the crop to market virtually impossible. Three months after the hurricane, the Pan-American Highway that links the nations of Central America was still underwater, and the ports of Corinto, Sandino, Arlen Siu, San Juan del Sur, San Carlos, Moyogalpa, San Jorge, and Cabezas were too damaged to handle regular shipping traffic.[24]

When First Lady Hillary Rodham Clinton visited Central America a few weeks after the hurricane, a Nicaraguan woman approached her to say that the hurricane "was worse than both the war and the earthquake that [had] leveled Managua." When Clinton expressed surprise, the woman explained that during the 1972 earthquake and the Contra war, "certain parts of the country were left untouched and life could go on somewhat normally." Hurricane Mitch, on the other hand, had affected the entire country. There was no place to which citizens could escape. Nicaragua needed more than just short-term humanitarian relief; it needed long-term reconstruction.[25]

Humanitarian aid workers encountered a similar situation in neighboring Honduras. The extraordinary rainfall caused by Hurricane Mitch caused rivers to overflow to a level the United Nations called "unprecedented in this century," with "very serious flooding on the coastal plain, landslides and avalanches in mountain slopes, and raging river rapids."[26] All of Honduras's eighteen territorial departments were affected in some way, but officials recorded the greatest devastation in the northern departments of Cortés and Colón, the southern region of Choluteca, and the capital city of Tegucigalpa.[27]

As in Nicaragua, unsustainable building, infrastructure, and agricultural practices made this both a policy-driven and a natural disaster, and the poor endured the most significant consequences. As one Honduran scholar noted, "The rains fell on all Honduras, but not everyone got drenched."[28] The settlements of the working poor, constructed of cheap or available materials such as adobe, palm, thatch, and metal and scrap, made their dwellings especially vulnerable to the high winds and flash floods. Entire communities washed away, including the village of Zacatales, which was "swallowed whole by floodwaters."[29] In the town of Morolica, near the border with Nicaragua, two rivers—the Texiguat and the Grande—converged and swept away close to 300 houses and public buildings. All the survivors had to be relocated, but only after the town's mayor was forced to walk 200 miles over the mountains to reach the capital city to request help.[30] Outside the commercial city of San Pedro Sula, the fastest growing city in Central America, the Chamelecón and Ulua Rivers overflowed and put a 120-square-mile area under fifty feet of water.[31] Thousands lived in shantytowns on the outskirts of the city, on deforested and unstable slopes, and these populations were the most tragically affected by the hurricane.[32] In the capital city of Tegucigalpa, home to over 1 million residents, the waters rose as high as the tops of telephone poles; Colonia Soto, one of the many hillside shantytowns, vanished under a landslide, and five other colonias were partially destroyed. The Laguna del Pescado dam, a natural dam and reservoir created years prior when authorities failed to remove a landslide on a tributary of the Choluteca River, collapsed and swelled the waters of the Choluteca, flooding much of the urban center of Tegucigalpa. The Concepción dam, also on the Choluteca River, was already full before the rains started; fearing that the dam would break, the dam operators released waters at dramatic speed, contributing to the flooding of communities below.[33]

According to one journalist who chronicled the destruction in Honduras, Mitch "rerouted rivers, blew the tops off mountains and in general rearranged the topography so much that sometimes engineers [felt they were] exploring virgin wilderness right in the middle of urban clusters."[34] The flooding, mud, damaged roads, and fissures several feet deep made towns and villages inaccessible except by boat or helicopter. When rescuers eventually arrived, they often found children tied to the tops of trees, having been strapped in place by their parents to prevent them from being carried away by the flash floods.[35]

The flooding and mudslides killed over 7,000 people in Honduras and injured more than 12,000; thousands more were unaccounted for and presumed dead. "In the immediate aftermath, there was a smell of death," said one Honduran. "Everyone in the capital wore masks."[36] Over 600,000 Hondurans were forced to move into temporary shelters in churches, schools, sports facilities, and tents. Three-quarters of the population are said to have lost family, property, or livelihoods.[37] The hurricane destroyed or damaged hospitals, medical clinics, and rural health outposts, as well as one-third of the country's public schools.[38] Power plants were down, telephone service interrupted, and sewer lines broken, dispersing sewage into the flood waters. The flooding also inundated warehouses full of pesticides and fertilizers that then seeped into the water, further endangering human and animal life.[39] Some 90 percent of the country's population had no potable water because of contamination, damaged reservoirs and purification plants, and impaired distribution.[40] Pools of standing contaminated water spread disease and exacerbated the public health emergency.[41] As in Nicaragua, the civil war–era land mines unearthed by the flooding and mudslides killed or wounded farmers trying to clear their fields of debris and dead livestock and prevented local governments from enacting major construction projects.[42]

Virtually all of the country's agriculture, industry, transportation, trade, and tourism were affected in some way. Land had to be cleared of debris, topsoil restored, crops replanted, equipment and machinery replaced, and buildings and infrastructure repaired, making a rapid economic recovery impossible in this developing country that already had a sizable debt before the hurricane. The hurricane perhaps did the greatest economic damage in the lowlands. Some 90 percent of the country's banana crop, one of Honduras's key agricultural exports, was destroyed. The two main banana companies—Chiquita Brands and Dole Foods, with headquarters in the United States—had joint losses of over $100 million, which subsequently resulted in the laying off of thousands of workers in Nicaragua and Honduras at all levels of production and shipping, just when workers most needed their salaries to rebuild. Not surprisingly, this prompted a public outcry, since, according to Malcolm Rodgers of the British relief agency Christian Aid, both companies stood to emerge from the disaster as winners because insurance payouts would "help modernize their operations and increase their global competitiveness." "It would have cost Chiquita just $6 million to pay their Honduran workers their $85-a-month wages for the year while [the company] rebuilt their plantations," said Rodgers.[43]

Since much of the coffee crop is grown on the highlands, only 15 percent of the crop was lost in the storm, but, as in Nicaragua, the condition of the country's roads, highways, bridges, and port facilities made shipping to market difficult. Coffee growers estimated losses at over $200 million. Honduran shrimp farms, another major source of income, were destroyed. Only the maquiladoras—the factories where textiles and electronic goods are assembled for overseas export (and the country's top source of income)—remained somewhat operational in the days and weeks after the hurricane. The *maquila* workers experienced fewer layoffs than workers in other industries, though many were left homeless after the disaster.[44]

Despite the region's susceptibility to hurricanes and other natural disasters, regional governments were not prepared to deal with the crisis, in part because of their poverty but also because they were post-conflict societies still dealing with the fallout of decades-long war. On November 2, overwhelmed by the "corpses everywhere," President Carlos Flores Facussé made an emotional appeal for international assistance: "We have before us a panorama of death, desolation and ruin throughout the national territory."[45] Some later accused the Flores administration of inflating the casualties to secure international aid (and by December, Honduran figures were adjusted), but the devastation was staggering nonetheless.[46] As Leopoldo Frade, the Episcopal bishop of Honduras, told friends in Miami, "Honduras has disappeared." The only way Americans could fathom this degree of ruin, he said, was to imagine every single U.S. state in shambles at the same time.[47]

In the immediate aftermath of the hurricane, President Flores Facussé declared a national state of emergency in his country, imposed a two-week curfew, suspended civil liberties, and ordered the armed forces to occupy key areas to prevent looting. But communities were left largely to their own devices to organize rescue operations and provide aid. As one anthropologist noted, Hondurans depended on their government to provide basic services and infrastructure, but communities were generally resourceful and self-reliant, accustomed to finding the help they needed for their own survival.[48] "For six months everyone was told to work on immediate disaster relief," said one Honduran university student. "No matter what your profession, education, or class, no one went to a regular job. Everyone worked in shelters, delivered food, or did anything that was needed. . . . It was a moment in our history of great solidarity."[49]

In the weeks and months following Mitch, civil society organizations across Honduras and other parts of hard-hit Central America regrouped to ensure that the international aid reached the neediest communities in a timely fashion, that decisions about reconstruction were conducted in a transparent and participatory manner, and that politicians and administrators at every level were held accountable.[50] One of the unexpected consequences of the disaster was "the intense national debate" over the following months on the need for transparency and a rebuilding of the country's democratic institutions.[51] "We are not just handing out rice and beans to people who are victims," said Noem de Espinoza, executive president of the Honduran Christian Council for Development. "We're working with leaders of rural communities, including the women and children, helping empower them to become subjects of their own destiny, and not simply objects of someone else's history. Hurricane Mitch gives us the rare opportunity to rebuild not just the physical infrastructure of the country, but also the human structures of power and decision-making."[52] Over the next few years, the United States Agency for International Development (USAID), in cooperation with a local nongovernmental organization, the Center for the Investigation and Promotion of Human Rights, funded "Transparency and Anti-Corruption" workshops to increase Honduran civil society's understanding of—and participation in—government processes, including the monitoring of the national budget.[53]

At the time of the hurricane, various international NGOs, including agencies of the United Nations, had an operational presence in Tegucigalpa, and these created an emergency coordinating group to work with people on the ground to assess damages, collect information, and mobilize personnel and other resources for relief and rescue operations. The United Nations channeled over $12 million in emergency assistance.[54] The UN Office for the Coordination of Humanitarian Affairs released regular "situation reports" to encourage international assistance. International aid arrived almost immediately, led largely by Mexico, the United States, and Spain, but eventually dozens of other nations participated, including Cuba, Taiwan, Sweden, and Germany.[55] International aid workers (including 200 Cuban medical doctors) arrived to assist with the distribution of meals, water, tents, and medical care—a daunting task given that so many bridges and roads had washed away. Because of the Honduran state's poor level of preparedness and concerns about the ability of state agencies and the military to administer aid in prompt and transparent fashion, the Flores administration assigned to Catholic and evangelical churches the task of receiving, managing, and distributing some of the foreign aid. These churches, in turn, worked with local faith-based and secular NGOs. Civil society organizations were especially determined that the military not divert the aid and engage in widespread profiteering, as it had during Hurricane Fifi in 1974.[56]

The Clinton administration monitored the situation in Central America carefully and provided Congress and the news media with regular updates on recovery efforts. In one White House press conference, U.S. national security advisor Samuel "Sandy" Berger underscored the strategic importance of assisting the affected countries: "This is a region that has been transformed in the past decade [after] searing civil wars claimed tens of thousands of lives in El Salvador and Nicaragua and Guatemala.... A disaster of this magnitude presents the region with a fundamental choice; it can undo the region's progress, or ... the countries of the region can work together to protect and even strengthen that progress [so that] over time, the region [can] emerge even stronger than before the storm."[57] In hearings before Congress, USAID officials assured legislators that any expenditures on relief and reconstruction served U.S. national interests: "As a neighbor, we should do what any neighbor does in a time of crisis: You lend a hand, you provide encouragement and you send a message of hope that puts this disaster behind us and opens the road to a better future for all of us."[58]

These compelling rationales for assistance sidestepped the fact that the United States had a particular obligation to the region. Central America's obstacles to recovery were due, in part, to the wars of the previous decades, which had exacerbated inequality and corruption and weakened political institutions and civil society organizations. The United States had played a role in creating these conditions because of its vast economic and military presence in the region. In Honduras alone, an estimated 60 percent of the Honduran economy was dominated by U.S. companies: the United States was the principal buyer of Honduran exports, served as the main supplier of Honduran imports, and became the biggest creditor for Honduran external debt.[59] But through its military presence and logistical support of particular political groups, the United States had also been a complicit actor in the politics of the region.[60] Honduras and Nicaragua were especially affected by this foreign policy since the United States had used Honduras as a base for launching a covert war against the Sandinista government in Nicaragua in the 1980s, which had forced both Central American countries to redirect economic resources toward their military and national security. The United States trained 15,000 Nicaraguan "Contra" soldiers on Honduran soil (as well as the Salvadoran soldiers used in the counterinsurgency campaigns), and this had left Honduras vulnerable to incursions from Sandinista forces.

A year before Mitch, the consul general of Honduras had reminded President Clinton that "the United States has a moral obligation with my fellow Central Americans. Some of them [have been forced to flee] for political reasons but others like the Hondurans [have fled] because of the economic problems that were created by the effects of such a war. This stopped foreign investment and caused an outflow of billions of dollars which increased unemployment and created more poverty."[61] The Clinton administration was conscious of this responsibility. In one press conference, First Lady Hillary Rodham Clinton remarked, "We understand that the future of the United States is linked to the future of the people of these countries."[62]

Also obscured in these discourses were the administration's concerns about another out-migration from Central America. The wars of the previous decade had generated hundreds of thousands of asylum-seekers in the 1980s, and their numbers had begun to ebb somewhat by the mid-1990s once the Contra war in Nicaragua had ended (1992) and peace plans in El Salvador and Guatemala were negotiated (1992 and 1996, respectively). A reinvigorated out-migration was now a possibility as countries struggled to rebuild their economies in the wake of the hurricane. This history of economic and military intervention and migration inevitably shaped American responses to Hurricane Mitch.

Disaster Relief as Foreign Policy

U.S. civilian and military personnel, including members of USAID's Disaster Assistance Response Team, arrived in Central America shortly after the worst of the storm had passed. There they partnered with local nongovernmental organizations and international relief workers to deliver aid to the communities left isolated by the storm. Relief efforts for the entire region were coordinated out of the Soto Cano military base in Honduras, fifty miles outside the capital and home to a regional task force of the U.S. Southern Command.[63] USAID-financed flights dropped desperately needed supplies in areas left isolated by the flood waters, but volunteers also used cars, boats, mules, and even hand-powered carts to deliver clean water, food, medical supplies, mosquito nets, home water purification systems, blankets, and plastic sheeting to the storm's victims.[64] Civilian and military personnel worked to remove pools of stagnant water and to repair water mains and sewer systems to prevent the spread of cholera, malaria, and other diseases. "With these and many other acts of humanitarianism," said one USAID official, "our government and its local partners helped millions of people survive those first terrible weeks."[65]

Soto Cano Air Base, Honduras, the hub of the international relief effort, February 1, 1999. (U.S. Air Force photo by S.Sgt. Gary Coppage)

Two weeks after the hurricane, Tipper Gore, the wife of the U.S. vice president Al Gore, led a presidential mission to Nicaragua and Honduras to personally assess the devastation. During November 9 and 10, Gore and her crew shoveled mud, distributed water and blankets, baked tortillas for families, and slept in tents alongside other displaced persons. "The people of Central America have suffered a disaster of Biblical proportions," said Gore. "The pace of their recovery depends, to a large measure, on what we as their neighbors do to help them, and on the long-term involvement of the international community." On her return to the United States, Gore's mission not only recommended a continued influx of emergency aid but also urged the Clinton administration to invest in the region's long-term recovery through debt relief, a temporary stay of deportation for unauthorized Central American immigrants in the United States, public-private investment partnerships, and the deployment of skilled Peace Corps volunteers to assist with rebuilding efforts.[66]

On the eve of her own visit to Central America (November 16–19), First Lady Hillary Rodham Clinton announced an expanded aid program to Honduras and Nicaragua.[67] To help Nicaraguan farmers recover more quickly from the erosion and the damage to agricultural infrastructure, USAID workers provided farmers with seed, seed-processing equipment, hoes, shovels, backpack sprayers, irrigation pumps, animal vaccines, and other supplies needed to return to normalcy.[68] To supplement government relief efforts, schools, churches, civil society organizations, and businesses across the United States donated food, clothing, and medical supplies; these materials were transported by U.S. military cargo planes and ships under the terms of the government's Denton program, which authorizes the use of military transport for humanitarian missions.[69]

By February 1999, the United States had invested over $300 million in emergency rescue and relief operations. U.S. personnel had repaired or built new homes, health clinics, schools, and rural roads; provided medical care, school supplies, farm equipment, and clean water and sanitation services; and conducted workshops to train local legislators and civil society organizations to plan and manage resources.[70] Members of the U.S. Naval Construction Battalion (the "Seabees") and various other engineering and road-building units assisted with the clearing and repairing of key roads, bridges, and other vital infrastructure or worked on the removal of the civil war–era land mines unearthed by the storm in Nicaragua and Honduras. Because Honduras had only four helicopters available for relief operations, over fifty U.S. military and Chinook heavy-lift helicopters arrived to assist with repairs and the removal of debris. At the height of the relief effort, 5,600 U.S. military and civilian personnel were on the ground in Central America, but 20,000 National Guard and Army Reserve forces also participated in some capacity, as did the Peace Corps' "Crisis Corps."[71]

This aid was not enough, however. On February 16, 1999, President Clinton requested from Congress a supplemental aid package of $956 million, not only to continue emergency relief projects in Central America but also to assist Haiti, the Dominican Republic, and other Caribbean nations suffering the effects of Hurricane Georges (September 1998), as well as Colombia, recovering from the earthquake of January 1999.[72] Of this package, $300 million was earmarked for Honduras and $94 million for Nicaragua. The additional funding brought the total U.S. assistance to these countries to more than $1.2 billion, an expenditure Brian Atwood of the USAID called "well worth it, given the amount of money that we've put in this region for other purposes in the past."[73]

While aid to other countries in Latin America was important, it was Central America that the Clinton administration most worried about. "Central America's full recovery from the storms is clearly in the U.S. national interest," said one White House memo: "Over the past decade, the region has made tremendous strides toward settling conflicts, strengthening democracy, promoting human rights, opening economies, and alleviating poverty. The economic destruction and dislocation caused by the hurricane threatens to undermine these achievements. Disaster assistance to our Central American partners will ensure that their transformation continues."[74] In their press briefings and congressional testimonies, administration officials assured an American public wary of more U.S. interventions in Central America that the United States was not carrying the financial burden of this recovery alone but was acting as part of a multinational campaign. "Assistance to Central America, although substantial, represents only an estimated 10 to 12 percent of the total needs created by the disaster," said one official. "We truly are part of an unprecedented worldwide effort."[75] In a February 1999 press conference, the First Lady pointed out to the American public that "this has been one of the most extraordinary international efforts in recent history."[76]

Despite these assurances that the U.S. was not responding unilaterally, the Clinton administration did play a prominent role in coordinating relief and recovery efforts and in negotiating debt relief, loans, trade benefits, and investment commitments from other governments and from the private sector. The administration called a meeting at the Inter-American Development Bank on December 9–11, 1998, attended by the presidents of Costa Rica, El Salvador, Nicaragua, Guatemala, and Honduras, as well as by representatives of the "Paris Club" nations, to discuss long-term debt relief and financial assistance to the region.[77] Those in attendance agreed that reconstruction required sustainable development designed to lessen ecological and social vulnerability, reduce foreign debt, achieve greater social equity, promote decentralization, increase civil society participation, and encourage good governance.[78] Central American governments were urged to draft reconstruction plans in consultation with civil society groups and to present their plans at a follow-up meeting in Stockholm five months later—a meeting to which civil society groups were also invited to participate.[79] Once plans were approved and in place, international donors would announce their longer-term financial commitment to the region. The World Bank (where the United States is the largest shareholder) secured $6.3 billion in guarantees from donor nations for emergency relief and reconstruction.[80]

These various negotiations and accommodations were considered necessary in a region with high levels of poverty and debt and where governments were regarded as authoritarian and corrupt. At the time of the hurricane, Nicaragua and Honduras had a combined foreign debt of roughly $10 billion and a combined under- and unemployment rate of close to 50 percent. Many donors considered Honduras a particularly risky investment because the government had defaulted on its loans in the late 1980s, and the World Bank, the International Monetary Fund, and the Inter-American Development Bank had declared the country ineligible for new loans. It was also regarded as a nation with high levels of corruption, cronyism, patronage politics, and inefficiency: in 1998, for example, the watchdog organization Transparency International had ranked Honduras one of the most corrupt countries in the world.[81]

But just as Mitch reenergized the civil society organizations in Central America that insisted on a rebuilding of democratic institutions, the disaster "rekindled a global debate on the morality of international debt, and whether poor nations stricken by natural disaster should receive special help."[82] The discussions at these various international meetings reflected the competing—and often oppositional—views about debt deferral, debt forgiveness, and the role of neoliberal reforms in long-term development. Several nations, including France and Cuba, chose to cancel all or part of Nicaragua's and Honduras's debts. By February 1999, Nicaragua's bilateral debt had been reduced by $46 million and Honduras's by $103 million.

But most of the debt relief came in the form of deferring the payments required to service the debt: the IMF/World Bank's "Highly Indebted Poor Countries" program, for example, established an emergency mechanism to temporarily freeze debt repayments so that countries could redirect the money they spent on servicing the debt to rebuilding national economies. The United States and other major creditor nations deferred bilateral debt service requirements for the next two years ($54 million in deferred payments).[83] The United States also contributed $25 million to the World Bank–managed Central American Emergency Trust Fund, adding to the $115 million pledged by nine other nations, to cover both countries' obligations to international financial institutions during the deferral period.[84]

These financial accommodations did not completely ease the debt burden, however; it only postponed the burden into the future (Honduras's financial obligation just to service its debt accounted for 40 percent of its national budget). The nongovernmental relief organization Oxfam called the deferral program "a stay of execution." The larger issues of why these nations were so indebted in the first place remained unaddressed. Critics pointed out that debt renegotiation usually required additional cuts to already decimated public services and resulted in the increased privatization of hospitals, public utilities, and natural resources. "The cancellation of unpayable debt, rather than a conditional moratorium, [would have been] a more just and effective tool to eradicate poverty and to achieve sustainable development and equity in the South," said one relief and development specialist. "If we continue to give with one hand and take away with the other, [countries are] likely to remain perpetually in debt to disaster."[85]

At the local level, much of the assistance also came in the form of loans: the Inter-American Development Bank and USAID, for example, offered grants and loans to 70,000 "micro-entrepreneurs" to help them rebuild workplaces, reestablish inventories, and generate employment.[86] The Clinton administration also secured a $200 million loan initiative from Citibank and the Overseas Private Investment Corporation to extend medium- and long-term loans to small businesses in Central America and the Caribbean. Recovery, then, required more indebtedness, at all levels of society.

With longer-term development in mind, the Clinton administration pushed for neoliberal "enhancements" to the Caribbean Basin Initiative (first established in 1983, during the Reagan administration). In the 1980s, the Caribbean Basin Initiative had encouraged agricultural diversification and the production of "nontraditional exports" to generate foreign exchange and stimulate economies. While these programs had increased the GDP in many countries, they had also increased land values and forced smaller producers off their land, ultimately increasing the number of landless farmers in the region who became part of the mobile workforce who lived in precarious conditions. Other economic policies, including the creation of industrial free trade zones and export processing zones in Honduras, crafted with the assistance of USAID, tried to redirect workers—especially female workers—into manufacturing plants in factory cities like San Pedro Sula with mixed results.[87] Despite these consequences for working people, and despite labor groups' opposition to privatization and free trade policies, the Clinton administration pushed for an "enhanced" Caribbean Basin Initiative, which the president hoped would increase production—and jobs—by making certain products (especially textiles manufactured in industrial parks and export processing zones) more competitive with Mexico, which benefited from the guarantees of the North American Free Trade Agreement.[88] Making Central American production more competitive with Mexico would stimulate the region's economic recovery and growth, argued administration officials, and would ultimately reduce migration to the United States.

As part of these neoliberal reforms, the administration offered duty-free and quota-free benefits to Central American countries that produced textile handicrafts or assembled textile and apparel products made from American fabric, yarn, or pieces cut in the United States. In order to be eligible for these benefits, countries had to demonstrate that they guaranteed workers' rights, established environmental protections, and cooperated on intellectual property issues.[89] Longer term, the administration was committed to the creation of a "Free Trade Area of the Americas" by the end of the century because, to quote one official, "if we are to help lift this region from crisis to recovery, we need to use the mechanisms of open markets that foster expanded trade."[90] The Flores administration supported these measures because it would assist Honduras's economic development and help stem migration to the United States; otherwise, Flores said, "there was little that [could] be done ... to reverse this [migration] trend."[91] (These economic initiatives, also pursued by the George W. Bush administration, culminated in the 2005 Central American Free Trade Agreement.)

Clinton administration officials hoped these various initiatives would lay the foundation for a more sustainable long-term growth that would minimize future political and economic crises.[92] "Our goal is a reconstruction that does more than replace what was blown away," Atwood told Congress. "We want to see the countries of this region achieve even stronger, more prosperous and more sustainable democratic development."[93] Atwood compared these investments to the post–World War II economic recovery program in Europe: "Fifty years ago, with the Marshall Plan and Point Four Program, our nation undertook a great and unprecedented experiment in strengthening nations less fortunate than ourselves. There have been many milestones in this undertaking, as diseases have been eliminated, life expectancies lengthened, democracies created, and nations lifted from receiving assistance to providing it to others." In the aftermath of Hurricane Mitch, said Atwood, "we have an opportunity to write a proud new chapter in our history—and to serve our own national interests as well."[94]

U.S.-funded initiatives to "restore and relaunch" Nicaragua and Honduras were based on three guiding principles: decentralization, accountability, and environmental security. Because "democracy [was] best constructed from the ground up," U.S. aid workers were told to work with local governments and civil society organizations, especially when designing mechanisms for evaluating accountability. Development projects in agriculture, transportation, and housing also had to be based on sound analysis, planning, and technological innovation so that investments would not be "washed away by the next natural disaster." The administration

President Bill Clinton addresses the people of Posoltega, Nicaragua, March 8, 1999. (Clinton Presidential Library)

insisted that U.S.-funded projects reach the traditionally marginalized sectors of society so they would not exacerbate long-standing economic disparities.[95]

On March 8, 1999, almost five months after Hurricane Mitch struck Central America, President Bill Clinton, accompanied by the First Lady, visited the sites of devastation in Nicaragua, Honduras, El Salvador, and Guatemala. The trip began in Posoltega, Nicaragua, where U.S. rescue and relief efforts had been most intensive. Clinton also met with Nicaraguan president Arnoldo Alemán. The following day, the president addressed relief workers at the Soto Cano military base in Honduras before traveling to the capital city of Tegucigalpa to meet with President Carlos Flores Facussé. The four-day trip concluded with visits to El Salvador and Guatemala.[96]

The state visit to Guatemala, to attend the Central America Summit in the city of Antigua, was the first by a U.S. president since Lyndon B. Johnson's short visit in 1968. That it took so long for a U.S. president to visit Central America was remarkable given the United States' long history of intervention in the region; Clinton recognized this when he admitted the visit was "long overdue."[97] During his stay, Clinton told the region's leaders,

"We cannot and will not allow a cruel act of nature to reverse the hard-earned progress of Central America." In the "Presidential Declaration" that followed, the eight presidents in attendance, including Clinton, committed themselves to work toward reconstruction, democratization and human rights, sustainable development, commercial investment, financial cooperation, and debt restructuring. On issues of migration, Clinton agreed to address the legal disparities created by previous immigration policies that assisted some Central American migrants to remain in the United States but not others. The eight presidents agreed to combat human trafficking, to ease the reintegration of those who willingly or coercively repatriated, and to create long-term employment opportunities that provided viable alternatives to migration. "We recognize that migrants contribute to the development and prosperity of their region of origin and to their new countries of residence," said the declaration, "but their migration must be orderly."[98]

At a meeting at the National Palace of Culture, President Clinton also apologized for the CIA's role in exacerbating the brutal thirty-six-year civil war in Guatemala that killed over 200,000 (over 80 percent of them Indigenous peoples). The apology came on the heels of the publication two weeks earlier of the findings of Guatemala's independent Commission for Historical Clarification (La Comisión para el Esclarecimineto Histórico), which examined the reasons for the war and its consequences. The commission conducted its research with the assistance of thousands of U.S. government records declassified and made available to them by the Clinton administration. "For the United States, it is important that I state clearly that support for military forces or intelligence units which engaged in violent and widespread repression of the kind described in the report was wrong," said the president, "and the United States must not repeat that mistake."[99] Unfortunately, subsequent administrations did not learn from that mistake. Perhaps most ironically, given the tone and content of Bill Clinton's speech, is that in 2009, former First Lady—and now secretary of state—Hillary Clinton gave tacit support to the coup that drove from office Manuel Zelaya, the democratically elected and progressive president of Honduras.[100]

In the years after Hurricane Mitch, concerns about financial mismanagement lingered, and Nicaraguans and Hondurans saw few of the billions promised by the international community. Sweden and Germany, for example, canceled their post-Mitch reconstruction projects in Nicaragua because these countries claimed that high-ranking government officials were siphoning off funds.[101] In Honduras, an estimated 165,000 citizens were newly plunged into poverty, and tens of thousands continued to live in "temporary" camps called *comunidades habitacionales de transicion*. Some of the homeless opted to leave these temporary camps and move into city shantytowns, which were now even more crowded, less safe and sanitary, and more vulnerable to extreme weather. In frustration, in May 2000, 700 families pushed past Honduran military troops and settled in an abandoned army base in the Bajo Aguán Valley, challenging the private developers who had purchased the land at bargain prices. These actions were part of a broader Indigenous movement to fight for access to land.[102] Over the next two decades, the Bajo Aguán Valley became the site of important environmental protests—and also some of the worst human rights abuses in Latin America, as a subsequent section in this chapter will discuss.

Central American Migration through a Complex Legal Landscape

The U.S. economic aid package was a humanitarian and political response to an environmental crisis, but one also designed to stem another out-migration from the region. A few weeks after the hurricane, Honduran president Carlos Flores Facussé issued a warning to North America that without substantial economic aid his fellow Hondurans would once again use migration as an adaptive response: "They'll walk, they'll swim, they'll run, but they'll go up north," he told the BBC.[103]

The Clinton administration knew Flores was right. For decades, Central Americans had coped with war, economic stagnation, and environmental disruptions through migration. Much of the migration had been internal. In Honduras, for example, rural-to-urban migration had escalated after the neoliberal reforms of the 1990s had decimated agricultural cooperatives and reconcentrated landholdings in the hands of a few individuals and corporations.[104] Pushed off of their land, farmers moved into the larger towns and cities where high rents forced them to settle on the outskirts, in dwellings on unstable hillsides and riverbanks that were particularly vulnerable to extreme weather. Others chose to migrate to neighboring countries in Central America or farther north to Mexico, the United States, and Canada.[105]

Prior to 1980, Central American migration to the United States had been modest compared with Mexican migration, but the wars had changed that dynamic, driving hundreds of thousands northward. By the late 1990s, a quarter of El Salvador's population lived outside the country. Los Angeles,

San Francisco, Miami, New Orleans, Toronto, Montreal, and Mexico City were among the North American cities with the largest Central American populations. As the peace accords were negotiated and signed, the war-driven migration ebbed and some Central Americans repatriated to their homelands, but environmental disasters reinvigorated internal and cross-border migration. In 1996, for example, Hurricane César left 100,000 people homeless in Nicaragua, and soon after, U.S. immigration officials reported a 50 percent increase in requests for green cards compared with the 1995 figure.[106]

Those displaced by Mitch followed these familiar patterns: before contemplating a move farther north, many first sought safety and economic opportunities within their own countries, moving from the harder-hit rural areas to urban centers or moving to adjacent countries. The countries in the region varied in their responses to the migration. One of the most generous responses came from Costa Rica. In December 1998, Costa Rican president Miguel Ángel Rodríguez announced a "Migratory Amnesty" that allowed unauthorized Central Americans to have legal permanent residency for up to two years with full access to education, health care, and other social services. This amnesty allowed 150,000 Nicaraguans who had moved to Costa Rica over the previous decades, as well as the more recent disaster refugees, to have some legal protections and avoid deportation. Migrants from Belize, El Salvador, Guatemala, Honduras, and Panama also qualified for this amnesty.[107] The four countries hit hardest by Mitch—Nicaragua, El Salvador, Honduras, and Guatemala—also had a migration agreement in place that allowed citizens to move about freely within the four countries for up to ninety days. All four countries reported increased migration from their neighbors; some of these migrants stayed permanently, while others transited through on their way to other countries.[108]

The Mexican government prepared for an increase in unauthorized migration, some of which would pass through the country on the way to the United States and Canada. The Mexican government had increased its policing of the Guatemala-Mexico border during the 1980s and early 1990s as a consequence of the wars in Central America, the Zapatista rebellion in Chiapas, and increased pressure from the United States.[109] Mexico did not criminalize unauthorized migration, but the government's migration services worked quickly to apprehend, detain, and deport unauthorized Central Americans. On the eve of Hurricane Mitch, Mexican authorities were deporting roughly 120,000 third-country nationals each year who, Clinton administration officials believed, entered Mexico in order to gain access to the United States.[110]

A Gallup survey prepared for the U.S. Information Agency reported that 600,000 adults in Nicaragua, Honduras, Guatemala, and El Salvador were considering or preparing to travel through Central America and Mexico to seek jobs in the United States.[111] Anticipating another large-scale migration from Central America, the U.S. news media increased discussion of the region. One *New York Times* exposé reported that most of the migrants heading to the United States were male heads of households who endured significant hardships along the journey: some were wounded or killed in accidents or as victims of criminal ambush; others were turned back at the Guatemala-Mexico or U.S.-Mexico borders. "If they catch me 1,000 times," one man was quoted, "I will try again 1,000 times until I succeed." Perhaps intentionally, the author of the exposé drew on the language of natural disaster, describing the migration from Central America as a "surge" of storm victims.[112]

In his testimony before Congress, Brian Atwood of USAID reminded legislators that "most [Central Americans] don't want to come to the U.S. but they are desperate to find work and provide for their families." Atwood urged Congress to support the president's economic agenda for Central America in order to stem this migration: "Our goal is to help those people stay in their native lands and return to the productive lives they led before the hurricane blew their world apart."[113] In a White House press briefing, Atwood reaffirmed this message and spoke of the urgency of continued economic investment in the region to convince Central Americans to stay home:

> Right now, people are trying to decide whether or not they are going to move to Costa Rica or the United States or whatever, or whether they think they might have a livelihood for themselves in their own country. If we continue to see the infrastructure damage, the roads out, the bridges out, people can't get their crops out to market, if we see small businesses destroyed, and don't make available to them credits so that they can resume their businesses, basically you're going to see further losses, because they will not realize their potential with respect to economic growth.

"That's why it's such an emergency," said Atwood. "That's why the United States needs to move fast."[114]

Clinton administration officials struggled to find an appropriate response to this new disaster-driven migration as well as to the hundreds of thousands of Central Americans already living in the United States without authorization. Almost as soon as he entered office, Clinton pushed for stronger deterrence, detention, and deportation policies to curtail unauthorized immigration as part of his broader national security agenda.[115] But long before Hurricane Mitch, administration officials understood that Central America desperately needed the billions of dollars its nationals sent home every year in remittances and consumer goods. They also understood that the mass deportations the administration had in mind would destabilize these fragile postwar societies struggling to achieve their democratic and development goals.

Congressional and White House interdepartmental memoranda in the years prior to Hurricane Mitch reveal sharp disagreements between those who wanted greater leniency toward unauthorized Central American migrants and those who felt it necessary to enforce immigration law. "[The Central American] countries have now made remarkable progress toward democracy, peace and reconciliation," said one advocate of increased deportation eighteen months before the hurricane. "Migrants no longer [face] danger at home and it is time to think of return—[the] U.S. cannot accommodate everyone who wishes to come here. Our task is to do this as humanely as possible and by working closely with countries of [the] region to minimize disruption to their economies and to the lives of migrants."[116]

The 1996 Illegal Immigration Reform and Immigrant Responsibility Act (IIRIRA) reflected this harsher stance. Prior to the IIRIRA, unauthorized immigrants could secure "suspension of deportation" if they demonstrated continual physical presence in the United States for seven years, were of good moral character, and could show that their removal would result in extreme hardship to themselves or to their citizen or permanent resident parents, spouses, or children. Those who qualified for suspension of deportation were allowed to work in the United States and to obtain permanent legal residency. IIRIRA raised the eligibility bar for suspension of deportation (now called "cancellation of removal") to at least ten years of physical presence in the United States. Those petitioning for cancellation of removal also had to demonstrate that removal would cause "exceptional and extremely unusual hardship" to a qualifying spouse, parent, or child, which the Board of Immigration Appeals interpreted to mean hardship that is "substantially beyond that which would be expected to result from the alien's deportation." IIRIRA also limited hardship-based cancellations of removal to no more than 4,000 cases in any fiscal year; and no more than 4,000 could adjust to permanent resident status in any fiscal year.[117]

IIRIRA had particular consequences for Nicaraguans because the law retroactively changed the manner in which their years of residence in the

United States could be calculated for purposes of establishing eligibility for cancellation of removal.[118] Also affected were the Central Americans at the center of the 1990 *American Baptist Church v. Thornburgh* settlement agreement (also known as the "ABC agreement").[119] Back in 1985, Salvadoran and Guatemalan plaintiffs had filed a class action suit against the Immigration and Naturalization Service, the Executive Office of Immigration Review, and the Department of State, charging hostile and discriminatory treatment in detention facilities and asylum hearings. The settlement agreement that followed granted the plaintiffs asylum hearings (or de novo asylum reviews if they had been removed from the United States without a fair hearing).[120] At the time of IIRIRA's passage, ABC asylum cases were still pending in the courts. Those who did not receive asylum often filed for suspension of deportation as a last-ditch effort to remain in the United States, but IIRIRA now imposed more stringent criteria for qualification, as well as an annual cap on those who could receive cancellation of removal in any given year.

Members of Congress complained to the Clinton administration that it was unfair to retroactively apply the IIRIRA provisions to their Central American constituents who were awaiting a resolution of their asylum cases.[121] Clinton administration officials agreed that several provisions of the law were "unduly harsh," "had gone too far," and were "inconsistent with the nation's humanitarian tradition" but reminded congressional critics that a longer-term remedy lay in the hands of Congress.[122] The administration approved one accommodation: until Congress passed new legislation, the INS would not deport individuals who would have otherwise qualified for suspension of deportation had it not been for the annual cap. The president also promised Central American leaders that there would be "no massive deportations and no targeting of Central Americans."[123]

Fearing that Hondurans might be excluded from any legal accommodation, the consul general of Honduras urged the president to grant a "five-year amnesty" to undocumented Hondurans in the United States, which he estimated at 1 million people (the INS placed the number of undocumented Hondurans at closer to 83,000). "The United States has now the lowest unemployment rate and the strongest economy," wrote the consul general. "These immigrants contribute a source of labor that is needed. Because of our countries' economic situation, we are not prepared to receive these fellow Central Americans at this time. Not only that we do not have jobs for them, but they also contribute to our small economies by sending dollars back to Central America." The National Security Council official tasked with responding to the letter only promised that the administration would transmit a legislative proposal that would "address the harsh and unintended consequences of our new law" for Central Americans.[124]

That legislation came in November 1997: the Nicaraguan Adjustment and Central American Relief Act (NACARA). The law allowed an estimated 200,000 Salvadorans and 50,000 Guatemalans (as well as certain nationals from Cuba and the former Warsaw Pact countries) to qualify for hardship relief from deportation, on a case-by-case basis, according to the more lenient pre-IIRIRA criteria. The law also allowed Salvadoran and Guatemalan petitioners to have their cases evaluated by trained asylum officers rather than by the immigration judges who normally preside over perfunctory and impersonal deportation proceedings.[125]

NACARA's principal beneficiaries, however, were the estimated 150,000 Nicaraguan and 5,000 Cuban nationals, most of them concentrated in Florida and California, who were allowed to adjust their status to permanent resident without meeting the hardship standard altogether.[126] Left out of consideration were Honduran nationals in the United States who, like Nicaraguans, had fled the disruptive Contra war of the 1980s. Hondurans staged protests in Miami and Washington, D.C., to call attention to their situation. In Miami, two Honduran Americans filed a lawsuit on behalf of 193 children born to Honduran parents who now faced deportation under the new law. "Thousands of children were born in the United States because their parents had to flee Honduras due to the war, danger, and chaos in their native country to a great extent caused by the United States," said the lawsuit.[127] President Clinton criticized NACARA for establishing differential treatment of Central American cases and pledged to work within the law to minimize the disparity and "to avoid a disruptive return [to Central America]."[128] Yet despite the president's concerns and the creation of legislative working groups to design greater parity, no significant adjustments were made to the law.[129] Over the next decade, Hondurans had the highest rates of deportation of any Central American population.[130]

This was the legal landscape that migrants who fled Mitch-torn Central America encountered in the United States. Those who had fled the civil wars, with more legitimate claims to refugee status, had waged a decades-long legal battle to remain in the United States with mixed results; those now fleeing environmental disaster found even fewer avenues for permanent residency in the United States. If they remained in their homelands and could afford the expensive application fees for an immigration visa to come to the United States, their chances for entry were not much better.

Immigrant visas were reserved largely for those who had skills necessary to the U.S. economy or those who had permanent resident/citizen spouses or children in the United States. Even if one met the criteria for immigrant status, a visa was neither guaranteed nor immediately received. The backlog for immigration visas was a long one, especially for countries in the Americas. Another track for admission was refugee status or asylum, but disaster-driven migrants could not qualify unless they demonstrated a "well-founded fear of persecution on account of race, religion, nationality, membership in a particular social group, or political opinion."[131] Because of these bureaucratic roadblocks, many of those who came to the United States in the wake of Hurricane Mitch chose to cross the border without visas (or overstayed their tourist visas), hoping to live and work anonymously in the service economies so that they could send remittances to their families back home.

In preparation for an increase in unauthorized migration from Central America, the INS drew up a contingency plan called the "Enhanced Border Control Operational Program" to open ten new detention centers that would house up to 5,000 people each.[132] Even though Central Americans transited through Mexico to reach the United States, there is little indication in declassified government records that the Clinton administration consulted Mexican officials when adjusting its immigration policies—and yet U.S. policies always affected Mexico in some way. When the Clinton administration decided to temporarily suspend deportations in the wake of the hurricane, for example, Mexican journalists reported a 30 percent increase in Central American border crossers in Mexico's southeastern sector, many of whom were hoping to take advantage of this new U.S. policy.[133]

Those living in the United States without authorization (or whose visas were set to expire) did have one legal avenue available to them: Temporary Protected Status, one of the provisions of the 1990 Immigration Act. According to this law, those unable to return to their countries of origin because of an ongoing armed conflict, an environmental disaster, or "extraordinary and temporary conditions" could remain and work in the United States until the Department of Justice (after 2003, the Department of Homeland Security) determined that it was safe to return home.[134] This provision, however, benefited only those who were already in the United States at the time of the political or environmental disaster, not those who fled in its wake.

For governments of disaster-struck countries to receive TPS designation on environmental grounds, the law required those governments to officially request TPS on behalf of its nationals living in or visiting the United States. (This was not required of countries experiencing armed conflict.) In the immediate aftermath of Hurricane Mitch, the Justice Department anticipated that requests for TPS were forthcoming, so the INS authorized a temporary stay of deportation for nationals of Nicaragua, Honduras, El Salvador, and Guatemala through November 23, 1998. On January 5, 1999, after receiving requests from Nicaragua and Honduras, Attorney General Janet Reno announced that nationals from these two countries were eligible for TPS due to "substantial disruption of living conditions" in their homelands. Nicaraguans and Hondurans could retain TPS for a period of eighteen months if they demonstrated continual residency in the United States since December 30, 1998.[135] By 2000, an estimated 100,000 Hondurans and 6,000 Nicaraguans had applied for TPS.[136]

The administration chose not to extend TPS to Salvadorans and Guatemalans. The official explanation was that Nicaragua and Honduras had experienced the greatest devastation and required the most sustained assistance. In the case of Guatemala, President Álvaro Arzu never officially requested TPS after Hurricane Mitch.[137] Over the next few months, Salvadoran officials continued to press for TPS. To have a large-scale deportation of their nationals now "would cause us tragic problems of stability," warned Salvadoran president Armando Calderón Sol. The INS extended the stay of deportation for Salvadorans and Guatemalans through March 8, 1999, the first day of Clinton's scheduled Central American visit, but the administration denied TPS, Deferred Enforced Departure, or a more extended stay of deportation.[138] "We must balance [our] significant acts of generosity against our legitimate interests in maintaining control of our borders," went the talking points crafted for the president's trip to Central America. "We believe that the future of all Central Americans is in their home countries and our commitment to relief and reconstruction intends to make this objective a reality." The administration projected that, by March 1999, El Salvador and Guatemala would be ready to absorb their deported nationals.[139]

After March 8, the Justice Department resumed the "limited and controlled removal" of the roughly 15,000 Salvadorans and Guatemalans in deportation proceedings, many of them having been reclassified as criminal aliens. (IIRIRA had reclassified thirty different offenses as "aggravated felonies" for which aliens could now be deported.)[140] Because deportations had to be negotiated with the home countries, however, "the pace of removals remain[ed] near a stand-still" throughout 1999. Declassified administration records suggest that U.S. officials did not push the issue too emphatically at first, fearing that they would "compound the economic problems of the Central American countries [if they deported] an overwhelming number of criminal aliens."[141] As First Lady Hillary Rodham Clinton said upon her return from Central America, "There is a double-edged problem here for those countries, with the immigrants. One is that they cannot accommodate people at this time. They've literally nowhere for people to live or be put. They also would collapse completely, in some cases, without the funds coming in from the people who are working and sending [remittances]."[142] These accommodations meant that those scheduled for deportation were held in detention facilities in the United States for a much longer period than had been the norm. In December 1998, 120 Central American detainees began a hunger strike at the Mira Loma Detention Center in California to protest their state of limbo.[143]

Clinton administration records also suggest that U.S. officials initially focused their resources on detaining and deporting criminal aliens—and not ordinary border crossers—in large part because of the financial expense associated with targeting a much broader population. To be sure, the "criminal alien" pool was now much larger than pre-IIRIRA because of the reclassification of certain legal infractions as felonies. Prior to IIRIRA, the term "aggravated felony" referred only to murder, federal drug trafficking, and illicit trafficking of certain firearms and destructive devices; the IIRIRA reclassifications now included lesser crimes or infractions such as simple battery, theft, filing a false tax return, and even failing to appear in court.[144]

The more hard-line officials in the administration advocated expanding the dragnet to include first-time border crossers as well, worried that any leniency might send the wrong message throughout the region. "[The] INS must have a credible deterrent along the border to send a message that the U.S. border is not open to illegal border crossers from Central America," said one internal memo. But the memo also recognized that, at current funding levels, the INS was unable to detain illegal border crossers from Central America, and they would have to be set free until their immigration court dates. "As word spreads that illegal aliens from Central America are not detained," the memo warned, "the currently manageable number of illegal border crossers may turn into a far larger and less manageable number of illegal entrants."[145]

Convinced by this argument, the Clinton administration requested from Congress a supplemental $80 million for fiscal year (FY) 1999 to pay for 2,245 detention beds in state and local facilities for criminal aliens, $20 million for 1,400 detention beds for noncriminal but unauthorized border crossers, and salaries for more INS detention staff.[146] This request, however, was significantly less than the 50,000 beds of the INS's original proposal. To discourage more Central Americans from undertaking the dangerous journey to the United States, U.S. embassies throughout the region used the Central American news media to disabuse would-be migrants that they would receive TPS or asylum if they made it past the U.S. border patrol.[147]

This information campaign, along with the local job creation that came with post-disaster reconstruction, may have stemmed Nicaraguan migration to the United States or redirected it elsewhere. In FY 1999, for example, only 873 Nicaraguans were apprehended along the southwest corridor, compared with 16,013 Hondurans, 7,569 Salvadorans, and 4,658 Guatemalans. Nicaraguans were also underrepresented among those deported: in FY 1999, only 397 criminal and noncriminal Nicaraguans were deported to their homeland, compared with 3,989 Salvadorans, 3,369 Guatemalans, and 3,349 Hondurans.[148] These interventions did not prevent out-migration altogether, however. Other countries in the region saw an increase in Nicaraguan migration, most notably Costa Rica, where Nicaraguans already made up 10 percent of the Costa Rican population of 3.7 million (in 1999).[149] Those who left Nicaragua in the wake of Mitch most likely settled in Costa Rica and other neighboring countries that were culturally familiar and geographically closer.

Hondurans, in turn, emigrated in ever-growing numbers. Prior to the 1990s, Honduran migration was largely internal rather than cross-border, but that pattern shifted, and an estimated 81 percent of those who left Honduras settled in the United States and Canada.[150] Of the U.S. cities with well-established Honduran populations, New Orleans was among the better known by migrants because it was a principal port of entry for the Honduran bananas imported by the Standard Fruit Company (now called Dole). In 1990, Hondurans numbered roughly 8,000 people in Louisiana (9.2 percent of Louisiana's foreign-born population): they were merchants, sailors, laborers, and workers in the food and service industries in or around the port city of New Orleans.[151] Migration to Louisiana, as well as to California, Texas, and other parts of the country, reached historic highs by 2005.[152] In New Orleans alone, the population was estimated at 140,000 to 150,000 by the time of Hurricane Katrina (August 2005), making it the largest Honduran population in the United States.[153] After Mitch, the Honduran migration was more socioeconomically diverse, consisting of not only the working poor but also members of the middle class who either entered without authorization or overstayed their tourist visas.[154]

Many of their compatriots never made it past the U.S.-Mexico border: the INS reported a 40 percent increase in apprehensions of unauthorized border crossers in the Southwest from FY 1998 to FY 1999 (from 20,814 to 29,115), 55 percent of whom were Honduran.[155] Even though Hondurans migrated in lower numbers than Salvadorans and Guatemalans, by 2002 more Hondurans were detained and deported than any other Central American group.[156] Despite these obstacles, in some parts of Honduras, migration became an important part of day-to-day life. As anthropologist Daniel Reichman found in his study of coffee farmers in the wake of Mitch, migration "was transformed from a relatively unusual practice of last resort to a routine part of social life . . . [used] to manage the cyclical nature of the [coffee] trade." Migrant remittances also became an increasingly critical response that "benefit[ed] the poor and inject[ed] badly needed *dolares* into the struggling economy."[157] As early as 2000, the Honduran government estimated that 25 percent of the population received "a substantial part of its income in the form of remittances from abroad," totaling $450 million.[158] By 2003, remittances to Honduras had grown to $860 million; and by 2005, remitted wages were Honduras's single largest source of foreign exchange, surpassing even the prosperous *maquila* sector. Remittances had increased the consumption capacity of the general public, allowing many to purchase or remodel homes, businesses, and other property. Millions more were sent home in appliances, computers, cell phones, and other consumer goods. As scholar Manuel Torres Calderón noted, migrants had become Honduras's most important export, more valuable than the country's banana and coffee crops; without their remittances, Honduras would be unable to pay its foreign debt.[159] In Nicaragua, remittances accounted for a quarter of the country's GNP.[160]

In deportation matters, the Honduran government generally cooperated with U.S. Justice Department officials to expedite the issuance of travel documents to facilitate the removal of Hondurans who had entered illegally into the United States. It also granted permission for the deportation flights of the Justice Prisoner and Alien Transport System to land in San Pedro Sula in addition to Tegucigalpa. But the Honduran government also requested financial assistance for reintegrating deportees, victims of trafficking, and returnees left stranded while in transit. Between 2014 and 2018, USAID provided $27 million to the International Organization for Migration to assist with the reintegration of returned migrants (especially unaccompanied children) in Honduras as well as in El Salvador and Guatemala. Most of the integration efforts focused on finding homes and jobs for the returnees and enrolling their children in school, but in limited cases, the organization also provided returnees with financial assistance to buy the material and equipment they needed to establish small businesses.¹⁶¹

Temporary Protected Status and Other Vulnerabilities

Temporary Protected Status was conceived as a short-term humanitarian accommodation of foreign nationals in the United States who were unable to return to their homelands because of political or environmental upheaval. The average TPS designation was eighteen months, and during this period, TPS recipients were permitted to seek employment and attend school. When set to expire, the government frequently extended the TPS coverage because conditions in home countries had not improved; consequently, some TPS recipients remained in the United States long enough to marry and create families, purchase homes, graduate from college, and establish successful careers and businesses. TPS did not grant permanent residency, however, and unless TPS holders found some other way to remain in the United States, they eventually had to return home or try to remain in the country without the protections of legal status.

By 2019, three appellate courts had offered mixed guidance on the issue of qualification for permanent residency. The Sixth and Ninth Circuit appellate courts ruled that TPS holders could adjust their status to permanent residence through family-based or employment-based petitions, even if they had entered the country without authorization. The Third and Eleventh Circuits, on the other hand, had ruled that TPS recipients who had entered the country without inspection were ineligible for adjustment of status. The position of the Department of Homeland Security, applicable in the other circuits, was that TPS holders were not eligible to secure permanent residence unless they departed the country to try to have a visa processed at a consular post; but for TPS holders who had entered the country without authorization, such an action could trigger bars to re-entry for up to ten years, according to the provisions of IIRIRA.¹⁶²

In 2021, the Supreme Court finally weighed in on the issue of whether TPS holders could apply for permanent residency. In the case known as *Sanchez v. Mayorkas*, the judges ruled unanimously that immigrants allowed to remain in the United States for humanitarian reasons could not adjust to permanent residency if they had entered the country unlawfully. The case was brought by José Sánchez and Sonia González, unauthorized immigrants from El Salvador who had received TPS after two earthquakes devastated their homeland in 2001 but were subsequently denied permission to adjust their status to permanent residency. The U.S. Court of

Appeals for the Third Circuit, in Philadelphia, had ruled against them on the grounds that immigration law requires applicants for permanent residency to have been "inspected and admitted" into the United States. Writing for the Supreme Court, Justice Elena Kagan agreed with the appeals court, stating that adjustment of status was limited to those who had been "inspected and admitted or paroled into the United States." "Because a grant of TPS does not come with a ticket of admission," she wrote, "it does not eliminate the disqualifying effect of an unlawful entry." Congress could have addressed this issue, said Kagan, by ruling that all TPS recipients held lawful admission.[163] (At the time of this writing, legislation introduced in the 117th Congress—the American Dream and Promise Act of 2021— included a provision lifting bars to permanent residency.)[164]

The congressional record does not reveal whether the architects of the 1990 law considered the possibility that some TPS holders might not be able to return home. This seemed to be the case for many Nicaraguans and Hondurans in the United States. Despite the influx of billions of dollars in humanitarian and development aid, conditions in Nicaragua and Honduras failed to improve. In the years after Hurricane Mitch, Central American countries continued to experience the economic dislocations that accompanied natural disaster, which made recovery, development, and economic growth difficult. Compounding these economic challenges were the extraordinary rates of political and criminal violence that emerged in Honduras, in particular, but also in Guatemala and El Salvador. By 2012, Honduras had the highest murder rate in the world: 85.5 murders per 100,000 people (by 2019, the number had dropped to 41 per 100,000).[165] Drug trafficking was one driver of the violence in the region, especially in Honduras, which became a transshipment point for narcotics trafficking through air, land, and maritime routes from South America to the United States.[166] Central American gang members, deported from the United States, engaged in extortion, human trafficking, drug smuggling, racketeering, and providing protection for the Mexican drug cartels.[167] But some of the violence was also the consequence of government and industry crackdowns on labor unions, lawyers, journalists, farmers, and environmental activists who vocally opposed privatization and other economic policies that fostered inequality. Corruption, the *mano dura* (iron fist) tactics of national security forces, the reemergence of private paramilitary squads, and the proliferation of arms (largely imported from the United States) all contributed to the political and economic destabilization.

The countries of Central America varied in their levels of institutional capacity and political will to confront these problems. The violence contributed to insecurity, but despite these conditions, it was poverty that Hondurans cited as the principal reason for leaving their country.[168] National economic policies pre- and post-disasters had raised the cost of living and made it difficult to secure a living wage. These realities made it unlikely that Central Americans could—or would want to—return to their homelands.

Honduran government officials and the Honduran news media pressed the White House to place TPS recipients on a permanent track toward citizenship and to authorize temporary work permits for their undocumented nationals in the United States. Editorials in San Pedro Sula's daily newspaper, *La Prensa*, offered what became the government's standard rationale: "The physical effects of that natural phenomenon can't be seen anymore, but its human and social consequences are still evident in most households and in the country in general, and that's why we need all the aid we can get from those who have left the country."[169] The members of the Central American Integration System also pressed the United States for some type of temporary work status for undocumented Central Americans who did not qualify for TPS; in the meantime, the organization worked to reduce the financial costs associated with the transfer for remittances to assist the region economically.[170]

Over the next two decades, the Honduran and Nicaraguan governments repeatedly requested extensions of TPS for their nationals on the grounds that their countries did not yet have the capacity to reincorporate a large number of repatriates.[171] Both Republican and Democratic administrations complied on the advice of their diplomatic corps. "While most USG-funded post-Mitch reconstruction efforts are complete, and much physical infrastructure has been rebuilt," said one 2004 cable from the U.S. embassy in Tegucigalpa, "the stagnant Honduran economy and the continued crisis situation in Honduran government finances make it unlikely the country [can] provide the jobs, health care, housing, and schooling for the approximately 87,000 people that might return if Temporary Protected Status (TPS) is ended."[172] But diplomatic officers also recommended that "Washington give consideration to the desired end game of TPS": "What was designed as a temporary program is now viewed by Hondurans as an annual process in which a decision not to extend TPS would be a shock. The eventual ending of the program, without some final clarification of TPS beneficiaries' immigration status, could put a large category of people into legal no-man's land and undermine USG [U.S. government] efforts to better control our borders." Whether through executive order or congressional legislation, the U.S. government should plan for an ordered end to

TPS, urged Larry Leon Palmer, the U.S. ambassador to Honduras. "The alternative, in which TPS eventually ends and 87,000 people are suddenly illegally in the U.S., could provide serious challenges to the USG [U.S. government]."[173]

Such a scenario also created legal challenges for the TPS beneficiaries. By 2020, the renewal of TPS without a pathway to citizenship had created a population of residents who were both part of the nation and yet legally on its margins: protected from removal for almost a quarter century but unable to secure permanent residency and citizenship unless they found other ways to legally achieve it.

By the second decade of the twenty-first century, Nicaraguan and Honduran TPS recipients had been joined by a new generation of Central Americans, many of them children, teens, and young adults, who made their way to the United States in search of safety and jobs. Some families felt they had no choice but to send their children unaccompanied to live with friends or relatives in the United States in order to save them from the forced recruitment, reprisals, and sexual assaults that were increasingly common in communities with a gang or paramilitary presence; others came to join parents who were already working in the United States. Their chances of receiving asylum or permanent residency were virtually nonexistent, but that did not stop the migration.[174]

In 2014, Central Americans apprehended trying to cross the southern border outnumbered Mexicans for the first time, and by 2016, there were an estimated 425,000 Hondurans and 70,000 Nicaraguans living in the United States without authorization.[175] The administrations of Bill Clinton, George W. Bush, and Barack Obama removed or returned undocumented foreign nationals in large numbers every year (a "removal" generally occurs through court orders, while "returns" are released across the U.S.-Mexico or U.S.-Canada border without a formal order of removal). Almost 28 million people (from different countries) were removed or returned during these three administrations.[176] Given these hard-line responses, it was all the more surprising that these three administrations continued to extend TPS for the Nicaraguans and Hondurans of Hurricane Mitch.

This safety net was threatened for the first time during the Trump administration, which tried to eliminate TPS protections as part of a comprehensive plan to overhaul the entire immigration system. Nationals of six of the ten countries that held TPS in 2018 found their status revoked despite warnings to the Trump administration from U.S. diplomats that several countries remained unstable.[177] Among the foreign nationals who stood to lose protected status were the Nicaraguans and Hondurans of

Hurricane Mitch. According to DHS secretary Kirstjen Nielsen, conditions in both countries no longer supported the status's continued designation.[178] The Salvadorans granted TPS in March 2001 also received news that their protected status would end.[179] Numbering 247,697, Salvadorans were the largest group of TPS holders in the United States in 2018, followed by an estimated 79,415 Hondurans. At 4,421, Nicaraguans were one of the smallest groups of TPS holders. The three Central American countries accounted for over 80 percent of all TPS recipients.[180]

Many Americans applauded the decision to terminate TPS, arguing that it had been much abused by both Republican and Democratic administrations. The Clinton, Bush, and Obama administrations had routinely extended TPS every eighteen months, they complained, so that tens of thousands now felt they had legitimate claims to permanent residency. Discussing Salvadoran TPS, one editorial in the conservative *Washington Times* complained that "the American Left have interpreted temporary to mean 'permanent' and [TPS's] protections ... an entitlement that dare not be revoked. But it isn't. Temporary means temporary":

> But one thing that hasn't been heard over the wails of sorrow from the New York Times ... is a simple and profound "thank you." Here's what it should sound like: "Thank you, America, for granting us temporary protective status back in 2001 when an earthquake struck in El Salvador when we were in the United States. Even though most of us had broken your immigration laws and entered your country illegally or over-stayed our visas when the earthquake happened. Nevertheless, your kind and generous president, George W. Bush, and the loving Republicans in Congress decided to grant us temporary refugee status because our home country had been devastated. We had no idea that the temporary status would extend for 17 years. It has been amazing and kind for you to allow us to stay all these years. And now that we have to go home 18 months from now, we'd just like you to know how grateful we are. Again, thank you."

"Wouldn't that be nice?" asked the editorialist. "We'd even grudgingly accept 'gracias' if you must, but since you've been in America for nearly two decades, you should, at this point, be able to bring it in English."[181]

After two decades in the United States, did the Nicaraguan and Honduran TPS holders have legitimate claims to permanent residency? Many advocates thought so. TPS recipients had established deep ties to their U.S. communities. According to one important study of Honduran (and

Salvadoran) TPS holders, the freedom from deportation had allowed them to have home-ownership and better job prospects, as well as health care and education.[182] They had raised their families here. Their impending removal now undermined everything they had accomplished in the United States and threatened to disrupt the homes, families, and communities they had created here. Salvadoran, Honduran, and Haitian TPS holders alone were parents to over 273,000 U.S.-citizen children who now had to leave the only country they called home if they wished to remain with their parents.[183]

TPS holders filed lawsuits to try to remain in the country. On October 3, 2018, in a decision known as *Ramos et al. v. Nielsen et al.*, the U.S. District Court for the Northern District of California enjoined the DHS from implementing and enforcing the decision to terminate TPS for four countries, including Nicaragua and El Salvador, pending further resolution of the various lawsuits making their way through the courts.[184] The Trump administration subsequently reversed course and extended TPS two more times, until October 4, 2021, for six of the ten countries currently qualifying for TPS designation—Nicaragua, Honduras, El Salvador, Haiti, Nepal, and Sudan. (All except Sudan had originally qualified because of environmental upheavals.) Immigration advocates pressured the Biden administration to redesignate Nicaragua, Honduras, and El Salvador for TPS—and now also include Guatemala—to expand eligibility for a wider range of people who had fled "natural disasters and dangerous and unstable conditions," but administration officials instead chose to grant only a fifteen-month extension to those who currently held TPS.[185]

For legislators on both sides of the ideological divide, the Nicaraguan and Honduran examples demonstrated the faults and limitations of Temporary Protected Status. Not surprisingly, during the 116th Congress alone (January 3, 2019, to January 3, 2021), over two dozen bills were introduced into the House or Senate to amend TPS in some way, but none was enacted.[186] The ideological divide in Congress was too deep to breach. None of the bills offered an acknowledgment of—or a viable response to—future climate-driven migration.

Lessons Learned, Warnings Ignored

Hurricane Mitch was one of six hurricanes to cause significant damage in Central America and the Caribbean in 1998. The first two decades of the twenty-first century brought more environmental disruption—hurricanes, flooding, mudslides, earthquakes, volcanic eruptions, and drought. Geography makes the isthmus vulnerable to natural disasters, but accelerated climate change increases this vulnerability and makes it more challenging for countries to plan for and recover from disasters.

Four years before Mitch, U.S. government agencies sponsored workshops in Costa Rica to help policy analysts, scientists, and engineers from seven Central American countries assess their nations' climate vulnerability. These workshops were part of the U.S. government's "Country Studies Program" that provided technical and financial support to developing countries so they could "develop inventories of their anthropogenic emissions of greenhouse gases, assess their vulnerabilities to climate change, and evaluate response strategies for mitigating and adapting to climate change."[187] It is difficult to assess whether this training had any impact on relief and recovery efforts after Mitch. In general, resiliency projects, in tandem with humanitarian assistance and foreign aid, do help countries rebuild sustainably, but these programs work best when disasters are followed by periods of relative calm. Back-to-back disasters make it difficult for populations to recover and prepare for the future. The Mitch-related reconstruction programs funded by the United States terminated by 2005, but subsequent environmental disruptions required that countries continually rebuild and reassess.

To make agriculture more sustainable, local communities across Central America have promoted the conservation of soil and water, the diversification of farming, and the use of native seeds and climate-adapted crops. At the state level, government agencies, working with international actors, have also experimented with different programs ad hoc. At the request of the Central American Commission on Environment and Development, for example, the USAID-NASA project SERVIR has provided "state-of-the-art, satellite-based Earth monitoring, imaging and mapping data, geospatial information, predictive models and science applications to help improve environmental decision-making" in Central America. SERVIR tools have forecast potential crises to help nations respond in a timely fashion. SERVIR has helped governments manage fishing during "red tide" events, for example, and has monitored and tracked cyanobacteria outbreaks, resulting in changes to water treatment practices.[188]

More difficult to forecast—or shield populations from—are the behaviors and policies that have made environmental crises far worse: governmental and corporate corruption, paramilitary and criminal violence, and shortsighted development policies, all of which have redirected economic resources away from the social safety net, politically destabilized the region, and made it impossible for families and communities to thrive.

These behaviors and policies have created a "chain reaction of social vulnerabilities" that have resulted in long-term environmental degradation, poverty, social inequality, population pressures, and escalating international debt.[189]

Hurricane Mitch made existing social vulnerabilities obvious. Mitch plunged thousands of people in Honduras into a "life-or-death struggle with deepening poverty" that has been difficult to overcome.[190] Five years after Mitch, 63 percent of the Honduran population lived below the poverty line; by 2018, despite registering the second highest economic growth rate in Central America, over 48 percent of the population still lived in poverty.[191] Nicaragua fared somewhat better. In 2005, roughly half the population lived in poverty, but by 2019, that percentage had dropped to 30 percent despite government corruption, the devastation caused by Hurricane Felix (2007), the international financial crisis of 2008–9, and the political and social unrest of 2018.[192]

In both countries, women, children, and the elderly have been the most affected, a fact reflected not only in the casualty rates but in the lingering trauma. In the aftermath of Hurricane Mitch, social agencies reported an increase in crime, domestic violence, sexual assaults, and child exploitation. The U.S. embassy in Tegucigalpa reported not only the "trafficking in persons of women/children for prostitution in the U.S., and children for commercial sexual exploitation in Central America," but also "serious problems with child labor in several industries."[193] Infant and child mortality remained high after the storm—thousands of children died each year because of malnutrition, poor housing, and preventable diseases.[194] School attendance dropped because children's labor became even more essential to the survival of families—and also because so many schools were destroyed.[195] "The vulnerability of families and communities has not declined since Mitch [because post-disaster efforts have] focused on emergency assistance and not on ongoing disaster prevention," said one Honduran NGO worker.[196] Consequently, the poor of Central America have suffered disproportionately because "poverty remain[s] the greatest moral and practical problem of our time," wrote Malcolm Rodgers, but eventually disasters will affect everyone in some way. "We walk away from it at our peril."[197]

For many Central Americans, then, internal and cross-border migration has been a means of coping with disaster, and migration will only increase as climate change accelerates. Changes in the hydrologic cycle are forecast to produce more storms, flooding, and drought, affecting not only the urban and coastal areas where most of Central America's population lives but also the region's rich agricultural zones.[198] The impacts of these natural disasters will worsen because of the ongoing deforestation, soil erosion, and water pollution that has resulted from poor urban and regional planning and because of the unregulated exploitation of natural resources.

Months before Hurricane Mitch struck the isthmus, a report issued by the Central American Commission on Environment and Development warned that deforestation was occurring at a dangerous rate across the region. Commercial logging and unsustainable agricultural practices were pushing the rural poor onto less desirable lands or into urban shantytowns. "Mitch [consequently] showed that Central America was sitting on a time bomb caused by environmental abuse," said one official.[199]

The urban and rural poor didn't need agency reports to tell them about these realities. They knew from lived experience that without the effective regulation of land and natural resources, the drafting and enforcement of zoning laws, the setting aside of new land in urban areas for sustainable housing, the designation of safe havens in every community, and the establishment of a wide-reaching communications network to facilitate early warning and evacuation, future storms would continue to create widespread havoc. The failure to enact sustained and effective policies was affecting Central America's agricultural, fishing, and industrial production; urban and rural infrastructure; sources of energy and water; and also the health and life chances of the region's populations.

If the first two decades of the new century brought more environmental disruption, it also launched a reinvigorated fight for environmental, social, and economic justice. Indigenous communities were central to that fight. On June 24, 2004, an estimated 5,000 Hondurans from all walks of life, traveling from four cardinal points across the country (Olancho, Siguatepeque, Danli, and Choluteca), converged in front of the Presidential Palace in Tegucigalpa to call for a halt to the unregulated, illegal, and clandestine logging by foreign and domestic interests that led to deforestation. Among the co-organizers of this "March for Life/Marcha por la Vida" was Father José Andrés Tamayo, a Salvadoran priest who headed the Environmental Movement of Olancho. Tamayo and the other march organizers insisted that there be greater community participation in the management and utilization of local natural resources; an audit of forestry resources by a qualified international firm, with the participation of community and nongovernmental organizations; and the creation of an interinstitutional commission to supervise the wood industry.[200] They also demanded that the government bring to justice those who, with the cooperation of law enforcement, had murdered their own. Among those murdered was twenty-three-year-old Carlos Arturo "Oscar" Reyes, the co-organizer of the 2003 March for Life, who was gunned down outside his home.[201]

The U.S. embassy in Tegucigalpa credited the March for Life with reenergizing a "committed environmental protection movement in the country." But American diplomats also noted that "if the marchers continue to press their agenda . . . they could run into violent opposition by the powerful economic interests who stand to lose the most by greater environmental protection enforcement and any changes/reform to the current system."[202] This was an understatement. Tamayo, who won the Goldman Environmental Prize in 2005, received numerous death threats and was finally expelled from Honduras after the 2009 coup ousted Manuel Zelaya, the democratically elected president committed to social reform. This coup—apparently conducted with U.S. support—represented "a full takeover by the country's ranching, trafficking, palm oil, and mining interests."[203] Over the next decade, hundreds of Hondurans were killed for protesting dams, mining, logging, unsustainable agricultural projects, and the hoarding of natural resources, making Honduras "the most dangerous country for community land advocates." The remains of farmers, journalists, lawyers, public prosecutors, community leaders, and activists—many of them Lenca, Garifuna, and other Indigenous and Afro-descendant peoples—were found tortured and dismembered. Indigenous communities were forcibly removed from their homes and lands. Among those assassinated was Berta Cáceres, founder of the National Council of Popular and Indigenous Organizations of Honduras and another Goldman Prize awardee, killed in 2016 for trying to stop the Agua Zarca dam on the Gualcarque River. In an interview, Bertha Oliva, founder of the Committee of Families of the Detained and the Disappeared of Honduras, noted that current conditions in Honduras were far worse than during the proxy wars of the 1980s "because the levels of impunity are much stronger and they constitute the policy of the state or, as we say, the debris of the state."[204]

Guilty or complicit in these assassinations were high political officeholders (or their family members) and military and government security officers, many of whom had financial stakes in the palm oil, mining, logging, and hydropower companies. In their quest for privatization and control, these individuals (and their companies) had wrested land away from small landowners for decades; when intimidation and coercion failed, company security guards (known as *sicarios*) tortured and killed those who got in their way and burned down entire communities. In 2012, the International Criminal Court charged Miguel Facussé Barjum, the uncle of President Carlos Flores Facussé and founder of the Dinant Corporation,

Honduras's largest producer of palm oil, with crimes against humanity for the bloody war his company waged against the farmers in the Bajo Aguán region.²⁰⁵ Not surprisingly, Facussé had been one of the key supporters of the 2009 coup that ousted Zelaya. The Committee of Families of the Detained and the Disappeared of Honduras continues to chronicle the casualties of the ongoing struggle to preserve the land and livelihoods of ordinary Hondurans.

Environmental and human right activists in Honduras (and elsewhere in Central America) have long been aware that development policies and practices, when pursued in unsustainable fashion, can worsen the impacts of natural disasters. The indiscriminate and unregulated logging, mining, and damming of natural resources have destroyed communities or left them vulnerable to flooding, mudslides, soil erosion, loss of biodiversity, and the chemical poisoning of water sources. According to environmental activists, climate-resilient development in Central America requires the large-scale restoration of the agricultural landscapes managed by small farmers; the restoration of ecosystems such as the riparian forests and mangroves that protect the water quality of rivers and streams; engineering interventions to prevent the landslides and flooding that affect urban areas; and the creation of community concessions that enlist Indigenous and local community stewardship of natural resources. Such interventions would create hundreds of thousands of jobs, strengthen partnerships with civil society organizations, and rebuild a resilient social infrastructure that would address inequities and allow individuals to exercise their right to stay home.²⁰⁶

Governments must be held accountable for the violence against environmental activists, but so too must the international institutions that turn a blind eye to such practices. Through all the violence, the United States continued to provide Honduras with military aid, "in violation of the 1997 Leahy Amendment forbidding military assistance to governments violating human rights."²⁰⁷ In 2016 alone, the year Cáceres was assassinated, the United States spent $100 million in bilateral aid to Honduras, millions of which went to the military and police forces that persecuted environmental activists. The U.S. embassy has also actively encouraged American investment in the extractive industries at war with those trying to protect communities, natural resources, and livelihoods.²⁰⁸ International institutions should demand, as the human rights group Global Witness has urged, not only the protection of human rights as preconditions to aid and investment in the region but also true sustainable development that meets the needs of the present without compromising the lives and livelihoods of future generations. Ironically, in Honduras, "green economy" programs such as the "Reducing Emissions from Deforestation and Forest Degradation program (REDD+)" have inadvertently rewarded some of the land barons accused of some of the most egregious human rights violations.[209]

Sustainable development—a development that "meets the needs of the present without compromising the ability of future generations to meet their own needs"—is critical to social justice and human rights.[210] The UN's Sustainable Development Goals, when implemented, strive to preserve communities and minimize the displacement that forces people to migrate internally and across national borders. Sustainable development can also help populations adapt to, and recover more quickly from, the upheaval that comes with accelerated climate change. Despite these efforts, however, migration, as a temporary or permanent response to disasters, will likely continue. And, as scholar Rodolfo Casillas has noted, the climate-driven Central Americans who now transit through Mexico are traveling through areas that have also been affected by climate change and are experiencing their own climate-related dislocations.[211]

Instead of blocking the migration of workers, nations in the region could expand the number of visas to allow migrants to work legally in other countries and send remittances to their families and communities back home. In the United States, Temporary Protected Status has served that purpose, but it is an imperfect response to environmental displacement, benefiting only those who are already in the United States. It was designed as a temporary humanitarian response to a onetime event such as an earthquake or hurricane from which nations would eventually recover, allowing their nationals to safely return home. But, as this chapter has demonstrated, some TPS holders might never be able to return home. At what point should they be placed on a pathway to full membership in U.S. society? And how will the United States respond to future environmental migrants who seek opportunities to work in the United States and to send remittances to their homelands but find their entrance blocked?

Back in 1998, the Clinton administration recognized that climate change played a role in Central America's destabilization. In her syndicated column, First Lady Hillary Rodham Clinton warned, "Climate experts predict that ferocious hurricanes like Mitch are typical of the disasters we can expect in the future as a result of global climate change."[212] Two decades later, during the record-breaking hurricane season of 2020—the most active Atlantic hurricane season on record to date—Hurricanes Eta and Iota once again caused severe flooding and landslides in Central America, affecting the lives of millions of people who were already suffering from other disasters, including the coronavirus pandemic. Like Hurricane Mitch twenty-two years earlier, the two hurricanes (which drove through Central America as Category 4 storms) devastated both rural and major metropolitan areas and washed away homes, schools, businesses, roads, and bridges; flooded agricultural fields; and damaged infrastructure. Over 4 million Central Americans were affected. Once again, citizens crowded into temporary shelters, only now the cramped conditions made the spreading of the COVID-19 virus an additional threat. Though aid workers arrived from all over the world to assist with relief and recovery efforts, many believed it would take a decade or more to recover from this new set of storms. As scholar Manuel Torres Calderón noted, in a short period of time, the hurricanes had become two more in a long list of tragedies.[213] Honduras, some said, had become ground zero for climate change in Latin America.[214] In a region continuing to suffer from poverty, debt, and other natural and human disasters, it's not surprising that Central Americans once again resorted to an age-old coping strategy—migration.

CHAPTER 3

What Protections and Benefits?

Coloniality and Citizenship in the U.S. Territories

> This year's hurricane season is already the most violent on record, and it will continue until the end of November. The season fits a pattern: changes to our climate are making extreme weather events more severe and frequent, pushing communities into a vicious cycle of shock and recovery.
>
> —UN secretary-general António Guterres, September 2017

> The sound of a storm of this magnitude is both indescribable and unforgettable. . . . To be devastated by a second Category 5 storm in 2 weeks is not something anyone should have to bear.
>
> —Congressional testimony, March 2018

On September 19–20, 2017, the Category 5 Hurricane Maria, the thirteenth named storm of the 2017 hurricane season, made landfall in the territories of the U.S. Virgin Islands (USVI) and Puerto Rico.[1] The two territories were already reeling from the impacts of another Category 5 hurricane, Irma, which just two weeks earlier, after causing total destruction in Barbuda, had passed north of the USVI and Puerto Rico before making landfall in Florida.[2] Even though the two territories

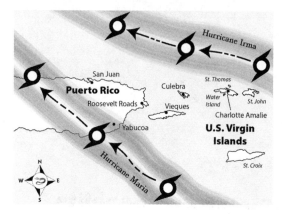

The U.S. Virgin Islands and Puerto Rico. (Map by John Wyatt Greenlee)

did not take a direct hit, Irma's 185 mph winds and thirty-foot surges killed eight people and caused over $3 billion in property damage. Over 1 million people in Puerto Rico and all of the USVI lost electricity, clean water, and communications. Residents were still dealing with floodwaters, landslides, mud, mold, debris, and downed power, phone, and cable lines when they received news that Hurricane Maria, another powerful storm that had caused "mind-boggling widespread devastation" in the Commonwealth of Dominica, was headed their way. This time, the eye of the 175 mph storm passed right through Puerto Rico, moving from the southeastern part of the island to the northwest. "It was as if a 50- to 60-mile-wide tornado raged across Puerto Rico, like a buzz saw," said one meteorologist at the National Center for Atmospheric Research.[3] Since 1851, only four storms at Category 4 or 5 had directly passed through the island, and Maria was by far the worst.[4]

As officials in Puerto Rico and the USVI realized that Maria was following so closely on the heels of Irma, they postponed recovery efforts and resumed emergency preparations. Hector Pesquera, Puerto Rico's public safety commissioner, urged the population to seek safety in one of the schools and public buildings designated as emergency shelters. He did not mince words. "You have to evacuate. Otherwise, you're going to die," said the commissioner. "I don't know how to make this any clearer."[5] Puerto Rico's governor Ricardo Rosselló warned the population of what lay ahead: "This is going to impact all of Puerto Rico with a force and violence that we haven't seen for several generations. We're going to lose a lot of infrastructure in Puerto Rico. We're going to have to rebuild."[6]

Prior to landfall, the U.S. secretary of defense directed the U.S. Northern Command—active-duty military and reserve officers, civilians, and contractors—to be ready to support any mission assignments generated by the Federal Emergency Management Agency.[7] Over 400 members of the USVI National Guard, called up in response to Irma, were joined by an additional 390 Guardsmen from around the United States. In Puerto Rico, 500 National Guardsmen were called up, and hundreds more waited at Homestead Air Reserve Base in south Florida for the relief and rescue operations to begin. On September 18, the day before the hurricane made landfall in the USVI, the Trump administration declared a state of emergency so federal funds could be released for disaster relief.[8] The airports closed and aircraft and nonessential government personnel relocated off the islands, but the U.S. Navy and Air National Guard kept ships, helicopters, and transport planes in the general area to assist with rescue operations.

Hurricane Irma had affected all of the Virgin Islands in some way, but St. Thomas, St. John, and Water Island suffered the most extensive damages, allowing St. Croix to serve as a hub for relief efforts. Now it was St. Croix, the largest of the islands, that was most affected. This second hurricane destroyed much of what remained standing. The islands were "drowned in almost 4 feet of rain in 12 hours," but the flooding was exacerbated because the infrastructure for drainage was still backed up with debris from Hurricane Irma. The beachfront hotels on which the tourism economy depended were severely damaged from the high winds and surges. Schools, hospitals, and police and fire stations were destroyed, as were homes, businesses, and government buildings. Landslides blocked key roads, making emergency services difficult. Even Hurricane Hole (Coral Bay), a protected cove on St. John where sailors and tour operators often moored their boats during big storms, was left in rubble after the back-to-back storms. Journalists reported that the islands' palm trees were snapped in half or stripped of their leaves, "leaving once lush tropical forests looking as if they were set afire with napalm."[9]

The hurricane continued its path westward and made landfall in Puerto Rico, 120 miles west of the USVI, at 6:15 a.m. on Wednesday, September 20. The storm entered south of Yabucoa harbor just short of Category 5 level with winds at 155 mph, but the storm was so powerful it destroyed the National Weather Service sensors and forced meteorologists to continue measuring its impacts by satellite. Thirty inches of rain fell in one day in some parts of the island (the amount of rain the city of Houston received over a three-day period during Hurricane Harvey a month earlier).[10] The hurricane destroyed Puerto Rico's outdated electrical infrastructure, which meant that over 3 million people lost not only electrical power but also access to clean running water, since the filtration systems required electricity. The storm knocked out 1,600 cell towers in addition to traditional phone lines, making communication impossible except by satellite phone. Roads and bridges washed away, impeding relief and recovery efforts, especially in the rural and mountainous communities. In the northwestern part of the island, the Guajataca Lake spilled over its banks, flooding communities downstream; fears that the ninety-year-old dam had cracked forced the evacuation of 70,000 people.[11]

On September 21, after the territorial governors officially requested assistance from the federal government, as required by law, the White House declared both territories disaster areas in order to authorize their access to a wider range of federal assistance programs.[12] San Juan's airport reopened to limited traffic, most of it military, and the U.S. Navy and Coast Guard partially reopened the city's major seaport to support daylight operations. By September 24, eleven ships had been able to dock in San Juan carrying water, cots, and generators for emergency power.[13] Brock Long, FEMA administrator, and Tom Bossert, of the Department of Homeland Security, among other Trump administration officials, made their first official visit to Puerto Rico on September 24 for a daylong assessment of the damages. By September 25, 2,600 Department of Defense personnel were in place in Puerto Rico and the USVI, including members of the USS *Kearsarge* amphibious ready group, which by the twenty-fifth had conducted eight medical evacuations and 123 airlifts and had delivered 22,200 pounds of relief cargo.[14] Members of the U.S. Army Corps of Engineers also arrived in Puerto Rico and the USVI to assist with "high-volume debris removal, infrastructure assessments, temporary emergency power restoration, temporary roofing and temporary housing."[15] The corps worked around the clock to stabilize the Guajataca dam, which, though not as damaged as initially feared, was pouring water from one section into the valley below. (The Guajataca dam, completed in 1927, was one of thirty-eight dams on the island designated by federal authorities as having "high hazard potential"; it had not been inspected since 2013.)[16] The corps also gave particular attention to the safety of ports and shipping channels in Puerto Rico, which would be critical not only to relief efforts but also to the economic recovery of both American territories, since many imports to the USVI first passed through Puerto Rico's ports.

By the end of the first week, federal officials had reopened eight airports and eight seaports in Puerto Rico for daylight operations. In order to ease pressure on San Juan's main airport, Luis Muñoz Marín International Airport, the Department of Defense began using the old airfield at the

U.S. Air Force C-17 Globemaster III aircraft from Travis Air Force Base, California, loaded with humanitarian aid, September 22, 2017. (U.S. Air Force photo by M.Sgt. Joseph Swafford)

Roosevelt Roads Naval Station (which had officially closed in 2004) as a supply staging area and as the logistical hub for rescue and relief operations. Only eleven of Puerto Rico's sixty-nine hospitals had power, so the Army Corps of Engineers provided generators to prevent post-storm casualties. In the USVI, the airports in St. Croix and St. Thomas reopened for relief operations, and its major seaports reopened to limited traffic. Engineers also restored power to St. Croix's main water production plant.[17]

What Back-to-Back Hurricanes Looked Like on the Ground

As impressive as the federal government's response was in official press releases and agency reports, relief and recovery efforts looked very different to those who were directly affected by the hurricanes. In the weeks that followed, it became clear that both U.S. territories necessitated a more sustained investment of human and financial resources if their populations were to recover, but their territorial status limited the assistance they could receive.

The fallout from the storms, which some called "apocalyptic," was far worse than originally assessed. The two hurricanes made uninhabitable half a million homes in Puerto Rico and the USVI. The storms hurt agricultural production in both territories, but Puerto Rico was especially affected, losing 80 percent of its crop value ($780 million in losses). Hardest hit were the plantain, banana, and coffee sectors of the agricultural economy, but poultry, dairy, and livestock farmers also lost animals, feed, and buildings.[18] Inspectors of the Puerto Rico Electric and Power Authority (PREPA) reported that while the generators on the island were operational after the hurricane, 80 percent of the outdated transmission system and 100 percent of the distribution system were damaged by the storm. Since the power grids in the territories were not connected to the fifty states, the population had to wait until the high voltage transmission lines were replaced. It wasn't until August 2018, almost a year after the hurricane, that the utility restored power to all but twenty-five customers on the main island, but Puerto Rico's offshore island municipalities, Culebra and Vieques, remained without power longer still.[19]

Puerto Ricans also lacked access to sanitary running water—a problem that had antedated Irma and Maria but was made worse after the hurricanes.[20] Maria caused a third of the sewage treatment plants on the island to be knocked out of commission, resulting in the seepage of sewage into the water system and the spread of bacterial infections such as leptospirosis. National Guardsmen installed and maintained water storage units, trying to target the remote rural areas; but even after the delivery of water was restored, water quality remained poor and required boiling for cooking and drinking.[21] Under these hardship conditions, businesses, banks, government offices, schools, and libraries remained shuttered for months, making it impossible for many citizens to earn an income, shop for necessities, or receive an education.

The federal government's estimates of losses and damages varied widely in the first days after the storm and occasionally contradicted information shared by Puerto Rican officials. Immediately after Hurricane Maria, for example, the Department of Defense reported that 44 percent of Puerto Rico's population lacked access to drinking water but later adjusted the figure to 55 percent (1.87 million people). Two months after the storm, however, Jenniffer González-Colón, Puerto Rico's resident commissioner in Washington, D.C., suggested the figures might be higher when she told Congress that more than 80 percent of the population still lacked access to either running water or communications.[22] The casualty rates were especially contested. The federal government's official death toll was sixteen at first, but the figure later increased to sixty-four (with an additional five deaths in the USVI). Puerto Ricans disputed these figures because they didn't take into account the post-storm-related casualties that resulted from the lack of food, water, life-saving medicines, and other services. As in other disasters, the poor and the elderly faced a higher risk of dying in the first six months after the storm. A George Washington University study released almost a year after the hurricane reported that 2,975 people died from storm-related causes in the first six months after the hurricane, and a Harvard study placed the number higher, at 4,645.[23] Journalist Sandra D. Rodríguez Cotto remarked, "Officials tried to hide the dead, but the stench of death prevented it. The dead were present, and their spirits demanded justice and respect.... To maintain their message that only sixty-four people died from the storm, when so many Puerto Ricans were burying their relatives, was a mockery."[24] For months after the hurricane, FEMA trailers parked outside the Forensic Sciences Institute housed decomposing corpses waiting to be claimed by loved ones for burial.

Most of these deaths occurred because of the lack of access to life-saving treatment in the wake of the hurricane: laboratories were unable to test for infectious diseases or environmental hazards; health centers and pharmacies were unable to supply enough vaccines, antibiotics, insulin, or even common anti-inflammatory drugs; and hospitals made errors because the lack of electrical power meant a lack of access to electronic health records.[25] The shortage of medical supplies was especially tragic given that Puerto Rico was a major producer of pharmaceutical drugs and medical devices. Indeed, damage to production plants on the island affected the delivery of intravenous saline bags and other life-saving medical supplies to the fifty states for over a year afterward.[26]

In the Virgin Islands, the restoration of electricity also occurred at a snail's pace; two months after the storm, over half the population remained without power.[27] Governor Kenneth E. Mapp estimated that repairing the power-generating plants and the transmission and distribution lines would cost $200 million, a major challenge for this small island state. The Virgin Islands' Territorial Emergency Management Agency, understaffed and underfunded before the storms, now struggled to provide emergency assistance across four different islands and called on the federal government for more assistance. Unfortunately, federal remediation efforts caused unnecessary frustration. FEMA's emergency stockpile, located in San Juan, had been quickly depleted after Hurricane Irma, requiring that humanitarian aid from the United States travel even longer distances to reach the USVI. Modular units for temporary schools and housing never arrived. In the first eight days, only forty-seven homes received tarps (under FEMA's "Blue Tarp Program") to shield homeowners and their families from the elements; and by early October, only 63 of the over 12,500 damaged homes had received semi-permanent roofing. Homeowners complained that the installation often produced more leaks in the already damaged roofs because these were designed for the shingle roofing found in Florida rather than for the metal "Caribbean construction" found in the USVI. Hotel rooms that could have accommodated those left homeless by the storm were taken up by the federal agency personnel who arrived on the island or were reserved for the tourists who never came that winter.[28]

As in Puerto Rico, the elderly of the USVI faced particular difficulties securing food, water, and assistance with debris removal. At a congressional hearing on disaster relief many months after Maria, a member of the USVI legislature used as an example his own difficulties securing assistance from FEMA and other government agencies: "[After Hurricane Irma] I initiated my FEMA application on the 14th of September. I was one of those few people who had internet access and I did the application online, the 14th of September. This is the 12th of March [2018] and except for the leaking tarp, except for the $500.00 from FEMA, I think that sums up about the assistance [I received]."[29] If this legislator, with all of his political connections and logistical advantages, encountered such obstacles, what frustrations did the average citizen endure? Despite the obvious devastation and repeated visits from inspectors and claims adjustors, FEMA often denied grants for home repair and rental assistance; the Small Business Administration denied loans; and private insurance companies routinely denied claims. As these stories circulated, many Virgin Islanders, expecting to be turned down, opted not to invest time filling out tedious forms to request assistance or compensation that would never come. In addition to the problems of housing, over a million cubic yards of trash and debris remained on public roads for months after the storm. Removing debris on the islands was always more expensive in the territories than in the continental United States—and it took months to determine where to relocate the debris and who would pay for it—but FEMA eventually assumed full financial responsibility for debris removal.[30]

The lack of adequate health care was especially acute in the USVI. Before the hurricane season of 2017, the USVI already suffered from a limited health infrastructure and a shrinking health care workforce. The hurricanes made these conditions worse, destroying the islands' two major hospitals in St. Croix and St. Thomas, as well as smaller clinics. Lost in the destruction was the USVI's sole cardiac catheterization laboratory and magnetic resonance imaging machine, as well as other diagnostic equipment and life-saving medications for diabetes, asthma, and heart disease. Once again, the most dramatically impacted by the disruptions in health care were the elderly and those with chronic or life-threatening illnesses who couldn't access health care or even receive packages of medical supplies from their stateside relatives because of the limited air traffic in the immediate aftermath of the hurricanes. The postal service was not fully functional, and the U.S. territories didn't qualify for the Denton program (the use of military planes to transport privately funded humanitarian aid), as Central America did after Hurricane Mitch.[31] To relieve the overworked doctors and nurses, wealthy citizens on the islands used their private jets to fly in relief personnel or to transport chronically ill patients to health care facilities in Atlanta or south Florida; 682 Virgin Islanders were evacuated for medical reasons, most of them critically ill, dialysis patients, or people in nursing homes.[32]

The Centers for Disease Control sent "health communicators" to USVI radio stations, churches, relief centers, and wherever else people gathered in order to provide information on how to stay healthy during this crisis.[33] In the months that followed, however, the number of storm-related casualties increased, and after the morgue filled to capacity, authorities had to ease the government's strict curfew so citizens could bury their loved ones. The hurricanes also "amplified behavioral and mental health issues . . . resulting in [a rise in] gunshot wounds, domestic violence, alcohol use disorders and depression." Authorities reported a significant spike in suicide attempts. "People lost their jobs," said one psychologist. "A lot of them lost their housing, their support network. The stressors in their lives magnified to such an extent that they just felt they couldn't cope anymore." A year after the hurricanes, six in ten Virgin Islanders continued to exhibit signs of post-traumatic stress disorder.[34]

According to a Kaiser Family Foundation study, the share of Virgin Islanders who lacked health insurance (30 percent) was higher than in the continental United States (12 percent), but despite these realities, federal Medicaid funds on the islands were limited, since, unlike in the fifty states, U.S. territories were subject to statutory caps.[35] Territorial status affected other forms of federal assistance, as well. More than three-quarters of the USVI's 105,000 residents qualified for varying amounts of federal housing assistance after the storm, but due to the higher cost of living on the islands, the FEMA-funded repairs, capped first at $20,000 per household (and eventually raised to $25,000), did not fully cover the construction costs of roof replacement, much less other structural damages. A year after the hurricanes, a sea of blue tarps still covered homes throughout the islands, and only half of tourist lodgings were open for business, hampering the tourism industry on which the islands depended for 60 percent of their GDP.

Making the Hurricanes' Impacts Known When No One Wanted to Listen

In Puerto Rico, the journalists, photographers, and editors of the Centro de Periodismo Investigativo, as well as the island's commercial news outlets, worked around the clock to provide information to their fellow citizens through whatever media platforms were functional on any given day. WAPA Radio, one of the 125 radio stations in Puerto Rico, and the only station to survive the hurricane's impacts, broadcast information across the island to those who had the means to hear it.[36] Teams of journalists from Telemundo and Univision, the major Spanish-language news outlets in the continental United States, were among the first to arrive in Puerto Rico to report on conditions for consumers of Spanish-language television and digital news across the United States, many of whom had family on the island. Univision provided its viewers with an online map with up-to-date information on damages in seventy-eight municipalities, as well as a Google Doc called "#PRActívate" to help Puerto Ricans in the fifty states connect with family and friends.[37] #PRActívate subsequently became the hashtag Puerto Ricans used to communicate information on Twitter and other forms of social media.

The English-language news media in the United States was slower to shine a light on conditions unfolding in the U.S. territories, which was not altogether surprising to the territories' residents, who were accustomed to the neglect. News from or about the territories rarely made it into the major news outlets. Public opinion polls in the wake of Maria revealed that roughly half of Americans did not know that Puerto Ricans were U.S. citizens.[38] From September 9 through October 10 (one week before the formation of the hurricane through one week after the storm became inactive), English-language media outlets in the continental United States ran 6,591 stories online about Hurricane Maria; by comparison, during a similar time span, news outlets published 19,214 stories online about Harvey and 17,338 on Irma, the two hurricanes that directly affected the continental United States.[39] The lack of media attention was especially surprising given that 5.5 million Puerto Ricans lived in the fifty states (compared to roughly 3.3 million on the island) and had families and friends on the island. The five political talk shows that aired the Sunday morning after Maria made landfall together produced "less than one minute of coverage... and three out of the five shows didn't mention Puerto Rico [or the Virgin Islands] at all."[40] Through social media, the National Association of Hispanic Journalists called the situation unacceptable and urged the national and local media "to cover [the] devastation in [Puerto Rico with] equal attention [and] responsibility that would be given anywhere in [the] U.S."[41] In the USVI, officials also asked that their situation not be ignored.

Over the following weeks, more and more journalists arrived in the territories to investigate recovery efforts on the ground. Equally important to disseminating information were the grassroots efforts of citizen activists who through their blogs and social media accounts alerted Americans to the desperate shortages in food, water, fuel, electricity, batteries, medical supplies, construction materials, and other basic necessities.[42] The professional journalists and citizen activists reported on the inefficiencies in relief delivery and the government waste (the photographs of millions of water bottles left undistributed on airport tarmacs for up to two years after Maria and of the warehouse full of undistributed food, water, cots, and baby formula revealed especially tragic examples of this waste and inefficiency).[43] They photographed the mud, debris, and trash that remained on the streets for months. They criticized the insurance companies that routinely denied claims, prompting over a thousand lawsuits.[44] They contested the official government casualty rates, but more importantly they reported on the resiliency of the Puerto Rican people, who organized themselves to clear the streets of debris and rubble, to ration and share their food and water, and to take care of the most vulnerable in their communities. This reportage helped push past the "compassion fatigue" that inevitably set in when the United States dealt with concurrent disasters. It also pressured federal authorities to do more. The public pressure, in turn, forced FEMA officials to reverse their January 2018 decision to end the distribution of food and water.[45] As the months passed and the U.S. territories once again faded from view, these local journalists and activists continued to report on conditions on the ground and increasingly turned their attention to the predatory venture capitalists (also known as vulture capitalists) who descended on the territories to aggressively acquire distressed properties at the lowest possible prices.[46]

Of the English-language journalists working in Puerto Rico, David Begnaud deserves special mention for his proactive documentation of recovery efforts through both traditional and social media. Begnaud told the stories of ordinary citizens and gave them a platform for publicly expressing their hopes, fears, concerns, grievances, and frustrations, prompting one grateful social media follower to nickname him "St. David of Begnaud, the patron saint of Puerto Ricans."[47] He was later honored as a *campeón puertorriqueño* (champion of the Puerto Rican people) at the 2018 Puerto Rican Day parade in New York City. He also became the recipient of the 2018 George Polk Award in Journalism for public service.[48]

The Virgin Islands received much less media attention than Puerto Rico, which critics later attributed to a variety of circumstances: its significantly smaller and majority Black population; Americans' fetishization of the Virgin Islands as a vacation destination and not a place where people lived, worked, and raised families; the territory's lack of true representation in Congress; and Americans' ignorance of—and apathy toward—U.S. territories in general.[49] All of these elements likely played some role in the lack of media coverage. As in Puerto Rico, local journalists performed a heroic job providing information to their fellow citizens, especially since they and their families were also personally affected by the storms. The staff of the *Virgin Islands Daily News* bunked in the newsroom and published a newspaper almost every day despite having no access to phone or internet in the first weeks after the storm. The paper's staff writers, photographers, and office personnel broke the government's curfew to investigate stories and to deliver the newspapers on foot. Their news coverage provided the population with practical information such as road closures, safety hazards, and commercial businesses that had reopened, but they also published the names of those missing and helped reunite families. When necessary, they dispelled rumors and took the local government to task for failing to respond rapidly to the most vulnerable in the community.[50] Also keeping Virgin Islanders informed was the team from the 82nd Airborne's Civil Authority Information Systems, which provided information to the public through loudspeakers and helped repair radio stations.[51]

As communications between the continental United States and its territories were reestablished and family members in the states were able to travel to the islands to assist with recovery, the "DiaspoRicans" (members of the Puerto Rican diaspora) shared more and more stories of survival on social and digital media.[52] Some of the stories focused on the human desperation—the interminably long lines at supermarkets, food banks, gas stations, and ATMs; the looting and spikes in crime and domestic violence; the post-storm casualties caused by the lack of pharmaceuticals and medical equipment; and even the emergency interment of deceased neighbors because funeral parlors were closed and morgues filled to capacity. But many more stories highlighted the resiliency of the people, the spirit of volunteerism, and the ways neighbors looked after each other. Neighbors shared their battery-powered radios and threaded power cables between homes so that those who had electricity could share power with those who did not. "Health brigades" of volunteer doctors, nurses, psychologists, and medical students knocked on doors to identify those who needed help.[53] Families took in elderly neighbors and those with special needs when hospitals, rehabilitation centers, and nursing homes became uninhabitable, and they cared for the animals left abandoned in shelters and sanctuaries. In the town of Adjuntas, the solar-powered offices of Casa Pueblo, a community-based environmental organization, became an energy refuge, distributing 14,000 solar lanterns, equipping ten homes with solar power for dialysis and small refrigerators for medication, and setting up a public satellite phone (and, eventually, a community-controlled solar-powered microgrid).[54] Readers learned that students, seeking normalcy, were returning to school as soon as possible, even if it meant studying in tents or hot, moldy classrooms with broken windows and equipment. One woman related a story of a father desperately searching for ice because his son's life-saving medicine required refrigeration. When the emergency was reported on the radio, "Within five minutes, the parking lot was full of people bringing ice for the father." "The storm brought out the best in people," said one citizen. "The multitude of acts of kindness got us through the long days."[55]

Celebrities, some with heritage connections to the islands, tried to draw public attention to the plight of Puerto Ricans and Virgin Islanders. The Spanish-born chef José Andrés traveled to Puerto Rico and organized 19,000 volunteers in twenty-five kitchens to provide 3.5 million meals to people left without food and electricity. His nonprofit, World Central Kitchen, later committed $4 million to his "Plow to Plate" initiative on the island to help food-related small businesses and smallholder farms "[sustainably] recover [from] natural disasters as well as increase their capacity for food growing, distribution and sales over the long-term."[56] The award-winning Broadway actor and producer Lin-Manuel Miranda raised millions for arts groups in Puerto Rico and, with the support of corporations, private foundations, and nonprofits, launched a five-year multimillion-dollar initiative to reinvigorate Puerto Rico's coffee production. Other celebrities organized benefit concerts and relief initiatives or worked with community organizations.[57]

In the Virgin Islands, country music performer Kenny Chesney, a part-time resident of St. John, chartered flights to fly in needed supplies and to transport islanders in need of immediate medical attention. His private foundation raised money for relief efforts in the U.S. and British Virgin Islands.[58] NBA basketball star Tim Duncan, of the San Antonio Spurs, also played a role in relief efforts. Duncan, who grew up on St. Croix, delivered 150,000 tons of food and supplies to his homeland after Hurricane Irma but became even more proactive in relief efforts after Hurricane Maria. As he explained in an interview: "We're back home [after Irma] feeling good about ourselves and we see something else forming on the horizon. No way. That was my initial reaction.... There's no way there's going to be another one to come ten days later, in succession, over the same path and then affect the same area.... And now we [had] to regroup... and refocus our efforts and expand them."[59] Using his celebrity and community connections, Duncan convinced corporations, charitable foundations, nongovernmental organizations, and even sports teams, players, and coaches to donate funds and supplies for USVI hurricane relief. Within a month, Duncan had raised $2.7 million specifically for the Virgin Islands, which he supplemented with his own personal donation of $1 million. Cargo planes transported food, tarps, and generators, which were distributed to over 17,000 people at relief centers throughout the islands. (When cargo planes were hard to find because they were delivering aid to Puerto Rico, the NBA's Memphis Grizzlies, who play at FedExForum, arranged for two MD-11 FedEx planes to deliver goods.) To focus attention on the USVI, which was receiving comparatively little attention in the news media, the notoriously private Duncan gave more interviews in two months than during the entire final season of his basketball career.[60]

The efforts of journalists, citizen activists, celebrities, artists, and athletes helped ensure that the territories remained in the news and in public consciousness in those critical first weeks. In order to overcome the limits and burdens of territorial status and the neglect of an apathetic—and sometimes punitive—White House, this sustained media attention was necessary for the ongoing recovery efforts.

The Limitations of Territorial Status When Dealing with Disasters

In the weeks and months following the hurricane, the Trump administration was criticized for its handling of relief and recovery efforts in the U.S. territories. The president at first seemed genuinely interested in assisting the victims of the impending storm. On September 19, a day after declaring a state of emergency, the president, a prolific user of the social media platform Twitter, tweeted, "Puerto Rico being hit hard by new monster

Hurricane. Be careful, our hearts are with you- will be there to help!" The next day, the president tweeted, "Governor @RicardoRossello- We are with you and the people of Puerto Rico. Stay safe! #PRStrong." On September 21, Trump told reporters that Puerto Rico was "obliterated" but that his administration would help the island rebuild "with great gusto." The president called Puerto Rican officials to pledge his support, and Vice President Mike Pence spoke by phone with Jenniffer González-Colón, Puerto Rico's resident commissioner in the U.S. House of Representatives, to assure her of the administration's support.[61] If administration officials called USVI officials, no mention was made of it in press conferences or social media.

In the weeks following the storm, contact between the White House and territorial officials became more erratic and, in the case of Puerto Rico, disrespectful.[62] The president, who had been in office less than a year, seemed distracted by other issues on the domestic front, namely, the Republican Senate race in Alabama, the legal challenges to his ban on immigration from Islamic countries (the "Muslim ban"), and even the antiracism protests of NFL football players. By comparison, the president had shown much more attention to the Florida and Texas victims of Hurricanes Harvey and Irma a month earlier. The president's public comments about the U.S. territories suggested that he was unfamiliar with the islands' geographic location and perhaps even their political relationship to the United States, at one point calling USVI governor Kenneth E. Mapp the "president" of the Virgin Islands. His remarks regarding Puerto Rican officials, especially San Juan's mayor, who criticized the slow pace of relief efforts, took a particularly ugly turn.

Aid responses also differed. Within six days of Hurricane Harvey's landfall in Texas (August 25), the U.S. Northern Command had deployed seventy-three helicopters, three C-130s, and eight pararescue teams to assist with search, rescue, and evacuation; and within nine days, over 30,000 employees of FEMA, the National Guard, and other federal agencies were on the ground in the affected areas of Texas and Louisiana, providing meals, water, cots, and blankets. During this short period, FEMA had approved $141.8 million in individual assistance to Harvey victims. Similarly, after Hurricane Irma struck Florida on September 10, more than 40,000 federal personnel, including 2,650 FEMA staff, were in place within four days to distribute necessary supplies.[63] By comparison, FEMA initially approved only $6.2 million for Hurricane Maria's victims. Despite the U.S. Northern Command's state of alert, troops were not pre-positioned before Maria, and rescue teams such as the amphibious ready group were not immediately available post-storm.[64] The number of federal personnel in

Puerto Rico and the USVI was also much smaller: nine days after Maria's landfall, 10,000 federal employees, including 1,972 FEMA personnel, were on the ground in Puerto Rico and the USVI. As a result of public pressure, the number of personnel eventually increased to 20,000, but it took forty-three days for FEMA to "ramp up" its response, compared with ten days for Texas.[65]

The disparities were attributable, in part, to the territories' smaller size but also to overstretched resources. By the late summer of 2017, FEMA was responding to 692 federally declared disasters and emergencies. Human and financial resources were stretched thin, and coffers for disaster recovery were close to depletion. California counties hard hit by wildfires—the most destructive in the history of the state—were also competing for the agency's attention during the early fall of 2017. The National Oceanic and Atmospheric Administration later calculated the damages from drought, flooding, wildfires, and hurricanes in the United States in 2017 at $306 billion.[66] Congress financed FEMA's Disaster Relief Fund based on forecasts of disaster activity, but the scale of the 2017 disasters was "unprecedented," said FEMA officials, and required Congress to pass three additional emergency appropriations bills that year—in September, October, and November—totaling $120 billion, so that FEMA and other agencies and departments could address the various environmental crises.[67] At a meeting of the House Appropriations Committee, FEMA administrator Brock Long pushed back against the critics of federal relief efforts in the territories: "My staff is tapped out," said Long. "They work around the clock and bust their rear ends every day to help those who are in need, and we're doing the best that we can do in trying to move as quick as we can." On television news shows, Long blamed the inadequate response in the territories on "logistics."[68]

A FEMA report published ten months after the hurricane noted that "the complex combination of hurricanes Harvey, Irma and Maria and the California wildfires," as well as Hurricanes Jose and Nate (September–October), had "tested the nation's ability to respond to and recover from multiple concurrent disasters." But mistakes were made, to be sure. At the time Maria struck the islands, FEMA's warehouse in Puerto Rico was nearly empty: its supplies had been rushed to the USVI two weeks earlier to deal with the fallout of Hurricane Irma and had not been replenished. The agency also underestimated how much food and water the islands would need after back-to-back hurricanes and how hard it would be to get additional supplies to the islands. According to the *New York Times*' investigation, an early draft of the FEMA report admitted to relying on a five-year-old plan for dealing with earthquakes and tsunamis.[69] Due to staffing shortages, the agency had to borrow workers from other agencies, many of whom lacked the training needed to deal with the consequences of natural catastrophes. The report later released to the public admitted to staffing shortages and "inefficiencies in program delivery" during 2017 but noted that FEMA's response to Puerto Rico and the USVI was the "longest sustained air mission of food and water delivery" in its history.[70]

The "inefficiencies in program delivery" were in part due to conditions at the local level. A week after landfall, more than 10,000 shipping containers of food, water, medicine, and supplies lay stranded in the port of San Juan because of road closures, a diesel fuel shortage, and a lack of truckers to transport the goods. In those critical first days, when communications were down and roads were blocked with debris, only 20 percent of truck drivers found the means to report to work.[71] But those with experience in disaster relief felt the administration could have done more to overcome these circumstances. One vocal critic was Lieutenant General Russel L. Honoré, the official credited with turning federal relief efforts around after Hurricane Katrina (2005). Honoré assessed relief efforts in Puerto Rico a week after the hurricane and criticized the Trump administration's "slow and small" response. More people, supplies, and equipment should have been sent in advance of the storm, as federal rules require, said Honoré, and the Department of Defense should have been given greater authority over the response, because only the military has the ability, through its "expeditionary logistics," to move supplies quickly across an ocean and onto an island when ports remain closed. Instead, the military was underutilized, he said, and relief efforts began four days too late. "The President has shown again he don't give a damn about poor people," Honoré told CNN in 2017. "He doesn't give a damn about people of color. And [that] SOB that rides around in Air Force One is denying services needed by the people of Puerto Rico. I hate to say it that way but there's no other way to say it." In another interview six months later, Honoré again complained, "We have the U.S. Army and the Marine Corps. We go anywhere, anytime we want in the world. And [in Puerto Rico] we didn't use those assets the way they should have been used."[72]

Other "inefficiencies in program delivery" were attributed to the vendors that FEMA contracted to deliver food and supplies to the territories. While most vendors fulfilled their contracts, some made costly mistakes. FEMA awarded one Atlanta-based company, with no significant experience in disaster relief, a $156 million contract to distribute 30 million "ready-to-eat" meals in Puerto Rico; the contract was subsequently canceled when officials discovered that only 50,000 meals (incorrectly packaged, at that) had been distributed during the period that 18.5 million should have been delivered.[73] FEMA contracts to three companies to deliver tarps and plastic sheeting were also canceled when the agency learned, months later, that two of those companies had not delivered any tarps at all and the third had produced tarps of such low quality that they failed inspections. Some 125,000 tarps were eventually delivered to the island but many months after citizens first needed them.

A *Frontline* exposé also found that a quarter of the staff whom FEMA hired to provide aid were untrained, unqualified, and unable to handle the daunting challenges they encountered on the ground.[74] Many did not speak Spanish. Volunteer aid workers complained that they rarely ran into FEMA representatives out in the field; when FEMA representatives did show up in a particular area, they often announced their presence by social media, but populations lacking electricity, internet, and phone service couldn't access that information. Volunteers complained that the supplies FEMA distributed in local communities were frequently inadequate or insufficient: those in need often stood in line for hours only to receive a small box of snack food to feed an entire family. "You know, all the citizens should be equal, but that's not what I'm seeing," said one volunteer nurse who spoke to Yxta Maya Murray. "They're not putting enough effort in Puerto Rico compared to the other states where there have been disasters."[75]

Much of this mismanagement was made public many months after the hurricane. In the immediate aftermath of the storm, it was the president's public comments and tweets that most contributed to the perception that the island territories were simply not a priority for his administration. In a series of tweets on the evening of September 25, the president (once again ignoring the Virgin Islands altogether) appeared to blame the inefficiencies in relief operations on conditions beyond federal control: "Texas & Florida are doing great but Puerto Rico, which was already suffering from broken infrastructure & massive debt, is in deep trouble," said the president.[76] Two days later, the president blamed the slow response on geography: "It's very tough because [Puerto Rico is] an island. In Texas, we can ship the trucks right out there, you know, we've got A-pluses on Texas and Florida and we will also on Puerto Rico, but the difference is this is an island sitting in the middle of an ocean, and it's a big ocean. A very big ocean." On September 30, in a series of eighteen tweets over an eight-hour period, the president complained about the "fake news" on federal efforts in Puerto Rico. He was especially critical of San Juan's mayor, Carmen Yulín Cruz: "The Mayor of San Juan, who was very complimentary only a few days ago, has now been told by the Democrats that you must be nasty to Trump. Such poor leadership ability by the Mayor of San Juan, and others in Puerto Rico, who are not able to get their workers to help. They want everything to be done for them when it should be a community effort. 10,000 Federal workers now on Island doing a fantastic job."[77] The Puerto Rican people and their leadership, said Trump, were responsible for the troubles they now encountered.

Acting DHS secretary Elaine Duke also contributed to the escalating frustration among commonwealth officials. On September 28, as the death toll climbed and desperate families complained that the emergency food, water, and supplies were failing to reach them, Duke told reporters that she was "very satisfied" with the government's response to the crisis: "I know it is really a good news story in terms of our ability to reach people and the limited number of deaths that have taken place in such a devastating hurricane." In response, Carmen Yulín Cruz told CNN, "Well maybe from where she's standing, it's a good news story." But "when you're drinking from a creek, it's not a good news story. When you don't have food for a baby, it's not a good news story. When you have to pull people down from buildings—I'm sorry, that really upsets me and frustrates me." At a press conference later that day the mayor said, "We are dying here and I cannot fathom the thought that the greatest nation in the world cannot figure out logistics for a small island. . . . So I am done being polite. I am done being politically correct. I am mad as hell. . . . If we don't get the food and the water into the people's hands, we are going to see something close to a genocide."[78] On September 29, the mayor, wearing a T-shirt saying "Help Us, We Are Dying," pleaded with American television viewers, "I will do what I never thought I was going to do. I am begging, begging anyone who can hear us to save us from dying. If anybody out there is listening to us, we are dying, and you are killing us with the inefficiency."[79]

Trump finally visited Puerto Rico on October 3. During his four-hour stay, the president once again provoked controversy. At a press conference at Muñiz Air National Guard Base, the president, clearly unaware of the climbing death toll, complained that Puerto Rico had "thrown our budget a little out of whack." He also insisted that Puerto Ricans were more fortunate than the victims of Hurricane Katrina, which had been a "real catastrophe." "Every death is a horror," said Trump, "but if you look at a real catastrophe like Katrina, and you look at the tremendous hundreds and hundreds and hundreds of people that died, and you look at what happened here and what is your death count? Sixteen people, versus in the thousands. You can be very proud. Sixteen versus literally thousands of

Governor Ricardo Rosselló, President Donald Trump, First Lady Melania Trump, and the governor's wife, Beatriz Areizaga Rosselló, at Muñiz Air National Guard Base, Puerto Rico, October 3, 2017. (U.S. Air National Guard Photo by Capt. Matt Murphy)

people." In a photo-op visit to a relief distribution center in the municipality of Guaynabo, outside of San Juan, Trump seemed to belittle the moment, tossing a half-dozen rolls of paper towels into the crowd. As he passed out flashlights, he told the crowd, "You don't need 'em any more, you don't need 'em," seemingly unaware that 90 percent of the population remained without electricity. Later that evening, wearing a T-shirt that said "Nasty" (a direct reference to Trump's earlier comments about her), Carmen Yulín Cruz told reporters, "This was a PR, 17-minute meeting. . . . In fact, this terrible and abominable view of him throwing paper towels, and throwing provisions at people, it's—really—it does not embody the American spirit."[80]

Though the president visited Texas twice during the first eight days after Hurricane Harvey, Trump chose not to visit the Virgin Islands at all. Instead, Governor Mapp was flown by military plane to meet with the president on the USS *Kearsarge* off the coast of Puerto Rico. Virgin

Islanders interviewed were not surprised that the president had bypassed the islands. "I never expected [the president] to actually come here," said one resident of St. Croix. "We're non-voters, what does he care?" Some expressed disappointment. "We deserve better. We are part of the states. Why would you not give us the decency and the uplift you would give us by your presence?" said another resident. Perhaps in response to the criticism, three days later Vice President Mike Pence, accompanied by his wife and Secretary of Transportation Elaine Chao, as well as USVI delegate Stacey Plaskett and Jenniffer González-Colón, visited the Virgin Islands for four hours; he flew over the sites of destruction and met with relief personnel before flying on to Puerto Rico to distribute food and pray with victims at a San Juan church. Pence assured Virgin Islanders that the Trump administration "stood with them" and was "here for the long haul." If Governor Mapp was disappointed with the pace of federal relief efforts, he chose to disguise the fact, assuring the Trump administration that Virgin Islanders were grateful for the assistance: "There is no country that responds to disasters like the United States of America," he said. Its efforts to restore power and operations were "down to a science."[81]

One unintended consequence of Trump's many gaffes was that the English-language media increased its coverage of Puerto Rico and, to a lesser extent, the Virgin Islands. As coverage and public pressure expanded, concerned legislators and politicians of both political parties urged a more proactive response from the federal government. New York congressman José Serrano, representing a district with many Puerto Rican constituents, urged the creation of a joint task force to oversee the rebuilding of Puerto Rico, similar to the task force created after Hurricane Sandy damaged parts of New York and New Jersey in 2012. Representative Adam Smith, the ranking Democrat on the House Armed Services Committee, pressed the president to establish a "coordinated military effort" similar to that created after Hurricane Katrina and Typhoon Haiyan in the Philippines (2013). Former secretary of state Hillary Rodham Clinton also called on the president and defense secretary James Mattis to increase the Pentagon's relief operations in Puerto Rico. Florida senator Marco Rubio, who had assessed the damages in Puerto Rico ten days before the president and warned of a "serious and growing humanitarian crisis," urged the president to offer the territories full federal support. Florida's other senator, Bill Nelson, also pleaded for more emergency aid: "In a crisis, all that matters is saving lives and giving people the resources they need to get back on their feet."[82]

Led by New York congresswoman Nydia Velázquez, eight members of Congress (six of them Puerto Rican or of other Latino ancestry) requested two regulatory changes to expedite the delivery of fuel, medicine, water, construction materials, and other essential supplies. In a letter to acting Homeland Security secretary Elaine Duke, these members of Congress asked the administration to waive the terms of the 1920 Jones Act (also known as the Merchant Marine Act) for one year. The Jones Act made relief and recovery difficult, they said, because the law requires that ships carrying goods between ports in the continental United States, Hawaii, Alaska, and Puerto Rico be built and registered in the United States, owned by U.S. companies (with at least a 75 percent American ownership), and manned by crews made up of at least 75 percent U.S. citizens. In a time of crisis, the Jones Act impeded the quick delivery of aid, especially if, as in 2017, there were concurrent environmental disasters that forced communities to compete for shipping and supplies. It also prevented the quick delivery of humanitarian aid from other countries. And because of the increased transportation costs, residents paid more for food, bottled water, and supplies—a hardship in ordinary times but a burden now when the population could least afford it. The administration had waived the Jones Act for seven days after Hurricanes Harvey and Irma to facilitate the delivery of fuel but stalled on a waiver after Hurricane Maria.[83] "We're thinking about [the Jones Act]," said Trump. "But we have a lot of shippers and a lot of people that work in the shipping industry that don't want the Jones Act lifted, and we have a lot of ships out there right now."[84]

The letter to Elaine Duke also requested a waiver of FEMA's required cost-sharing agreements. (Emergency assistance to the Virgin Islands and Puerto Rico, including debris removal, was provided at 75 percent federal funding and required the territories to make up the balance.) "Puerto Rico's current economic conditions have already pushed the local government's financial resources to the breaking point," said the letter, making it impossible for the territory's government to match any costs. In urging these regulatory changes, the members of Congress reminded the president that "the people of Puerto Rico have long been denied the same benefits provided to other American citizens. Today the stakes are just too high."[85]

Stacey Plaskett, the USVI's nonvoting delegate in the U.S. House of Representatives, and Governor Kenneth Mapp also pushed for certain accommodations. The federal government instituted a $20-million-per-fiscal-year cap on emergency funds allocated collectively to the territories of Guam, American Samoa, the Commonwealth of the Mariana Islands, and the USVI; by the time Hurricane Irma had struck the USVI, $15 million of that $20 million fund had already been spent addressing other disasters in the other territories. The fifty states were not subject to such a cap, and Governor Mapp argued that American citizens should be treated the same everywhere.[86] Plaskett also pointed out that Puerto Rico's larger size and population garnered it most of the resources and attention. "[The USVI] is the only place in the United States that has been hit by two Category 5 hurricanes—ever—within a ten-day span," said Plaskett. "And we don't hear about us in the news. Are we not Americans? Do we not serve in the military? Do we not do the things that all Americans do?"[87]

In March 2018, Plaskett hosted a formal hearing of the members of the U.S. House Subcommittee on the Interior, Energy and Environment at the USVI legislature in Charlotte Amalie, so Congress members could assess the plodding recovery efforts in the islands. This was the first time in U.S. history that the House of Representatives had held a formal hearing in the USVI, but only two members of the subcommittee bothered to travel to the territory to attend—Plaskett and Blake Farenthold of Texas, the chair of the subcommittee. Ironically, the chamber where the committee met lost electrical power during the middle of the session, requiring a temporary recess.[88] Throughout the hearing, the Virgin Islanders who testified repeatedly assured Farenthold that they were "grateful" and "appreciative" for the assistance received, but they also took the government to task for its half-hearted response and its role in creating additional hardships.

In these and other appeals, the territorial legislators reminded the administration that Puerto Ricans and Virgin Islanders were U.S. citizens, a framing some critics found problematic since it implied that citizenship—and not their shared humanity—should be the primary reason for assistance. In their appeals for aid, the territories' governors also resorted to this framing, which perhaps was necessary given that many Americans knew so little about their fellow citizens in these small island states. Puerto Rico's governor Ricardo Rosselló, for example, urged a rapid and comprehensive intervention "to prevent a humanitarian crisis occurring in America. Puerto Rico is part of the United States," he reminded the Trump administration. "We need to take swift action." Likewise, Governor Kenneth Mapp urged the federal government to remember "the forgotten Americans" in the Virgin Islands. "We are no different than Americans anywhere else," he said. In his November 2017 testimony before the Senate, Mapp reminded Representative Farenthold that Virgin Islanders were U.S. citizens and thus "fellow Americans."[89]

During the final months of 2017 and through 2018, federal agencies did increase their operations in these U.S. territories. The Pentagon sent additional cargo aircraft with relief supplies for Puerto Rico and the USVI. It dispatched the USNS *Comfort* to Puerto Rico, as Hillary Clinton and others had urged, and appointed Jeffrey Buchanan, a three-star army general, to oversee the military's response. President Trump also bowed to pressure and waived the Jones Act for a ten-day period. Eleven regional staging areas were distributing food and water to the population. Governor Rosselló announced that every resource he had asked for was either on the island or on its way. Most of the population remained without electricity or running water and thousands remained in shelters, but personnel were actively clearing roads, delivering supplies, and reopening hospitals. John Rabin, the regional FEMA administrator, assured the population, "Everybody here is in support of our fellow citizens in Puerto Rico. Our cooperation and collaboration with the entire federal family—the Department of Defense, Health and Human Services, the Department of Energy—all that is on display."[90]

Yet despite the expanding relief and recovery operations and the supplemental congressional authorizations of aid in 2018, the assistance was not enough to place these two debt-stricken territories on the path to rapid and sustainable recovery. At every turn, the hurricanes' impacts underscored the territories' unequal relationship to the fifty states and reinvigorated a century-old debate about coloniality and the rights, duties, and benefits of citizenship.

The Coloniality and Citizenship of the "Forgotten Americans"

In the face of catastrophic environmental disasters, political and economic ties to major powers like the United States or Great Britain should theoretically help citizens of territories adapt to—and bounce back from—their various impacts. But as the cases of Monserrat, Puerto Rico, and the USVI show, an appropriate response is always dependent on the political will not just of locally elected legislators but of legislators far away who control the economic and political lives and destinies of their territorial citizens.

Americans know very little about the territories acquired in their name. While Puerto Rico may receive a few mentions in U.S. history textbooks, usually in discussions of the War of 1898, other territories—the USVI, Northern Marianas, Guam, and American Samoa (and eleven uninhabited territories claimed by the United States)—rarely do. As travel writer Doug Mack noted in *Slate*,

> Crucially, the way we talk about the territories—and the fact that we rarely talk about them—goes hand in hand with their colonial

status. This, after all, is the very core of colonialism: treating people not as equal members of society but as abstractions, policies to be debated, problems to be solved. This isn't unique to the territories (ask residents of Flint, Michigan, or Ferguson, Missouri, or the Navajo Nation if they think their stories are well-told), but it's amplified in these islands; their residents are held not just in low regard but as entirely inscrutable, "foreign in a domestic sense," as the Supreme Court put it in one of the Insular Cases.[91]

The histories of Puerto Rico and the USVI illustrate how citizenship and political "association" with the United States do not yield the benefits and protections that Americans on the "mainland" assume. Consequently, as Yarimar Bonilla and Marisol LeBrón have pointed out, the legacies of colonialism set the stage for Maria's impact and aftermath: they established a "coloniality of disaster."[92]

The islands' territorial status is the product of the United States' "age of empire." Since the early nineteenth century, the United States had acknowledged interest in acquiring Puerto Rico, the Dutch West Indies, and other Caribbean territories not only for purposes of trade, commerce, and national security but also as symbols of the nation's growing influence and power.[93] U.S. corporate interests lobbied for their acquisition to secure raw materials for consumption and to expand the number of overseas markets. Puerto Rico, in particular, acquired strategic importance once U.S. policy makers began planning a canal across the Central American isthmus. Military strategists hoped to establish a permanent naval base on the island to guard the Mona Passage, one of the routes linking the Caribbean to the Atlantic.

The island became the spoils of the War of 1898. Article IX of the Treaty of Paris authorized Congress to determine the inhabitants' civil rights and political status, but it would take half a century to work out the details. A military government ruled the island until passage of the Foraker Act (First Organic Act of 1900), which granted Puerto Ricans the right to elect representatives to a House of Delegates; but the governor, executive council, cabinet, and justices of the Supreme Court were all appointed by the White House. In a dramatic departure from precedent, Puerto Rico was not placed on the path to statehood as other territories had been; and as many scholars have noted, the new administrative arrangement gave Puerto Ricans less self-determination than Spain's 1897 Charter of Autonomy.[94] A startling symbol of this unequal relationship is that congressional publications misspelled Puerto Rico for three decades.[95]

WHAT PROTECTIONS AND BENEFITS?

A series of U.S. Supreme Court decisions known as the "insular cases" (1901–22) affirmed Puerto Rico's unique status as an "unincorporated territory" belonging to—but not part of—the United States and subject to the "plenary power" under the Territorial Clause of the Constitution.[96] In the words of Justice Edward Douglass White, Puerto Rico was "foreign to the United States in a domestic sense."[97] As a result, this "unincorporated territory" became "a new legal and political reality." Legal scholar Efrén Rivera Ramos noted, "If before those cases were decided there were no such things as 'unincorporated territories,' after their rendition, the American political and constitutional world would include such entities. Through this doctrine, the United States was also constituting itself, officially, as an imperial world power. This, to my mind, is one of the most perfect examples of the law's 'power of naming' and of the capacity of the law to generate new understandings and therefore, new realities."[98] Judge Juan R. Torruella, in turn, wrote that the insular cases established Puerto Rico as "separate and unequal." Puerto Rico Supreme Court justice Luis F. Estrella Martínez called it "apartheid in the Caribbean."[99]

Until passage of the Jones Act of 1917, Puerto Ricans were "nationals" but not "citizens" of the United States. After 1917, citizenship became one of the "ideological pillars" of Puerto Rico's permanent association with the United States, according to Jorge Duany, but it was a citizenship that lacked the full constitutional freedoms, rights, and privileges of citizens in the states.[100] "Detaching citizenship from the right of political participation, as in the case of Puerto Rico" wrote Rivera Ramos, became "a central feature of the legal framework of the American colonial enterprise."[101]

The Americans appointed to government office in Puerto Rico and to the Bureau of Insular Affairs (1898–1939), which oversaw territorial affairs, represented the military or corporate interests of the United States, and their primary concerns were to extract wealth and repatriate it back to the mainland. Indeed, many did not bother to learn Spanish or to establish permanent residence. By 1934, every sugarcane farm in Puerto Rico belonged to one of forty-one syndicates, 80 percent of which were U.S.-owned.[102] Americans oversaw or controlled the postal system, the railroads and seaports, the public utilities, fruit growing and tobacco manufacturing, and shipping and banking. Workers displaced from their small landholdings either worked on these agricultural estates or enterprises or joined the migrant stream into Puerto Rico's larger towns and cities or to places as far away as New York and Hawaii. By the 1930s, the cost of imports, almost all of which came from the United States, were as much as 15 to 20 percent higher than in the states because Puerto Ricans were denied the right to enact price-fixing legislation.[103] This was especially devastating during periods of economic recession.

Over the next century, the limitations of unincorporated status became clearer. The Jones Act established an elected bicameral government modeled after the U.S. Congress, but all laws passed by the Puerto Rican legislature were potentially subject to annulment or amendment by the U.S. Congress. The Jones Act also gave Puerto Ricans the right to elect a resident commissioner to the U.S. Congress, but these resident commissioners had limited powers: they could introduce legislation and vote only in committees, not in the full House.[104] Puerto Ricans could not elect their own governor or vote for the president who sent them to war (Puerto Ricans have served in every war since the First World War). When the island legislature first tried to enact a minimum wage law to mirror the one in the United States, the U.S. Supreme Court declared it unconstitutional.[105] For a brief period, from 1948 to 1957, even the possession and display of Puerto Rican flags was illegal. Presumably in the interest of public health and the advancement of science, thousands of women and men were subjected to medical procedures, oftentimes without their full understanding or consent. (A 1968 study revealed that over 35 percent of women on the island had been sterilized.)[106] Under these conditions, it is not surprising that a robust nationalist or pro-independence movement emerged on the island in the first decades of the twentieth century (and continued into the 1970s) and was subject to intense surveillance by the FBI and local law enforcement agencies.[107] Struggles for greater self-determination were always hard fought and concessions rarely and reluctantly given.

During the Second World War, Puerto Rico became an important military enclave for the army, navy, and air force. Roosevelt Roads in Ceiba became the largest U.S. naval base outside the continental United States, and the offshore island municipality of Vieques became a site for military exercises. Unlike other countries around the world that decolonized in the decades after the war, Puerto Rico found its political, economic, and military ties to the United States strengthened. In 1948 Congress granted Puerto Ricans the right to elect their own governor, and in 1950 President Truman authorized a constitutional convention that would define the U.S.–Puerto Rico relationship "in the nature of a compact."[108] The new constitution, approved by referendum (with 76.4 percent approval), established that Puerto Rico was a commonwealth (or *estado libre asociado*, as it was called in Spanish). The shift in policy was designed to remove Puerto Rico from the United Nations' list of "non-self-governing territories" subject to the supervision of the UN Decolonization Committee, but in practice, commonwealth status did not fundamentally alter Puerto Rico's relationship to the United States.[109] Jorge Duany noted that while "commonwealth status represented a greater degree of autonomy for Puerto Rico in local matters, such as elections, taxation, economic development, education, health, housing, culture, and language, the U.S. federal government remained in control of most state affairs, including citizenship, immigration, customs, defense, currency, transportation, communications, foreign trade, and diplomacy."[110] In the decades that followed, Puerto Ricans were exempted from federal income taxes unless they worked for the U.S. federal government, but they paid payroll taxes for Social Security, Medicare, and unemployment, and the Puerto Rican government levied its own high taxes. The population qualified for veterans' benefits and some U.S. social welfare programs such as subsidized public housing and nutrition assistance (for example, food stamps), but not at the same levels applied in the fifty states.

Puerto Rico is, to quote Puerto Rican jurist José Trías Monge, "the oldest colony in the world." Trías Monge does not use the term "colony" "in the traditional pejorative sense of a people ruthlessly exploited by an evil empire," for he believes that the "compact" with the United States has yielded benefits. Instead, he has used the term to denote "that the United States unnecessarily holds excessive powers over Puerto Rico, thoughtlessly preventing it from attaining a respected place on earth as a self-governing people freely associated to the United States, integrated to the United States as a state thereof, or separate from the United States as an independent nation."[111] Others are much more critical than Trías Monge, insisting that Puerto Rico has functioned as colonies generally do: that is, to generate wealth or other advantages for the powers that administer them, in this case, the United States. Historically, the wealth generated on the island has been extracted and exported. In recent years, for example, over $35 billion in manufacturing profits, or approximately a third of the GNP, has been repatriated back to the mainland. U.S. manufacturing has produced several contaminated superfund sites on the island that need cleanup. Even the U.S. military presence, though a source of jobs and upward mobility for many, has come at a cost. Until 2003, when public pressure finally forced it to stop, the navy used parts of Vieques for live bombings and other military exercises that were disruptive to daily living, the local economy, and public safety, for example. When the navy eventually closed Roosevelt Roads, it took hundreds of jobs with it (in 2011, parts of the former base became a training site for the U.S. Army Reserve).

Dissatisfied with the political status quo, an overwhelming majority of Puerto Rican voters have consistently supported either statehood or a more sovereign or "enhanced" commonwealth arrangement.[112] Those who support the commonwealth association believe that it has brought security and a higher standard of living than is found in other Caribbean countries, but they would like Puerto Ricans to exercise greater control over their political and economic affairs. Most Americans in the fifty states regard Puerto Rico (and other territories) as somewhat of a mystery—an island over a thousand miles away with a foreign population despite a shared citizenship. Demographic and economic data for Puerto Rico, the Virgin Islands, and the other territories is hard to come by because such data is collected and reported less frequently than in the states.[113] Americans are perhaps more familiar with the larger Puerto Rican diaspora of 5.5 million people, once concentrated in New York and Chicago but now dispersed across the fifty states. Ironically, members of the diaspora exercise more political rights in the states than in the homeland: if Puerto Ricans relocate to one of the fifty states, they have the same rights and privileges as other U.S. citizens, including the right to vote in presidential elections. Full citizenship is dependent on what patch of earth one calls home.

The physical and economic devastation that Puerto Rico experienced in the wake of Hurricanes Irma and Maria was not just the result of the forces of nature. Puerto Rico's conditions were decades in the making. As journalist Sandra D. Rodríguez Cotto wrote, "People, not the storm, created much of the chaos. Puerto Rico's infrastructure had been abandoned for decades amid the island's recession and local partisan politics, and people died because the government failed to follow emergency plans established by previous administrations. Adding to all that was the racism and neglect of the Trump administration, which provided aid too slowly, and the corruption and ineptitude of the local government. This was a true recipe for disaster."[114]

The territory's public debt, taken on to cope with commercial and tax policies imposed by Washington and Wall Street, had caused disruptive economic upheaval that only worsened after the hurricanes destroyed factories, farms, businesses, and public utilities. The roots of the debt lay in the economic policies of the 1990s. To lure more businesses to the island, section 936 of the U.S. tax code, the "Possession Tax Credit," approved by Congress in 1976, had offered partial or full tax exemption to U.S. companies operating in Puerto Rico and other U.S. territories. As a result, the island had become an important site for the production of pharmaceuticals, electronic equipment, and scientific instruments; and because section 936 required companies to invest their tax-exempt revenues in Puerto Rico for at least six months before repatriation, Puerto Rico also developed an important financial and banking sector. All this began to change in 1996, when Congress passed the Small Business Job Protection Act that initiated a ten-year phase-out of tax breaks, with devastating economic consequences for Puerto Rico.

The new century, then, brought an economic freefall. The corporate tax breaks expired in 2005, but, anticipating these changes, many companies had already begun to move elsewhere, taking with them billions of dollars in bank deposits and tens of thousands of jobs. Puerto Rico (and the USVI) lost industrial competitiveness in a region now dominated by free trade agreements. By 2005, Puerto Rico was in an economic recession that was only aggravated by the global financial crisis of 2008. The administration of Luis Fortuño, representing the pro-statehood Partido Nuevo Progresista (New Progressive Party), declared a fiscal state of emergency and enacted a "stabilization plan" to try to stimulate economic growth: with the support of the Puerto Rican legislature, the administration cut spending by firing 30,000 public employees; raised tuition at public colleges and universities; created tax havens for high-net-worth individuals whom they hoped to lure to the island to invest and create jobs; and changed environmental protection laws to open public lands to private development.[115] The administration also secured billions of dollars in financial assistance through the Obama administration's American Recovery and Reinvestment Act.[116] When these policies failed to produce sufficient investment, revenue, and jobs, the administration took on a debt burden of $17 billion (22 percent of the GNP) in bonds and lines of credit, which subsequent administrations increased and were unable to repay or refinance. More public sector jobs were cut and taxes raised, leading to an impossibly high cost of living that drove more and more young and mid-career people to migrate to the continental United States. Between 2000 and 2015, Puerto Rico lost 9 percent of its population through migration (334,000), with the San Juan metropolitan area losing the most people (40,000).[117] By 2015, 45 percent of those who remained on the island lived in poverty though the jobless rate was 12.6 percent. In 2016, the median household income was $20,078.[118]

In 2014 three bond credit rating agencies downgraded Puerto Rican bonds to non-investment speculative grade, also known as "junk bonds." The owners and managers of hedge funds (and the more speculative "vulture funds"), many of them now taking advantage of the new tax haven, had purchased a fifth of Puerto Rico's debt at discounted rates and now pushed for the acceleration of debt repayments, even if it meant more government spending cuts, higher taxes, the reduction of the minimum wage, the firing of teachers, and other austerity measures.[119] These measures would be severe in any country with massive debt, but they were especially hard in U.S. territories where "a substantial amount of the wealth produced and profits derived in the island was repatriated and not reinvested in the country."[120] Vulture funds, in particular, bought billions of dollars of bonds at reduced prices, banking on the idea that Puerto Rico, as an unincorporated territory, would never have full bankruptcy protection.

In 2015, for the first time in its history, the commonwealth defaulted on $58 million in principal and interest due on Public Finance Corporation bonds; and the following year, the territorial government defaulted on almost $2 billion in constitutionally guaranteed "general obligation bonds." Declaring the debt unpayable, the government tried to declare bankruptcy but was prevented from doing so. The Court of Appeals for the First Circuit ruled that the 1984 amendment to the Bankruptcy Code preempted Puerto Rico from authorizing its municipalities and public utilities to declare bankruptcy; and in June 2016, the U.S. Supreme Court affirmed the decision 5–2 in *Puerto Rico v. Franklin California Tax-Free Trust et al.*[121] Congress's plenary power over Puerto Rico prohibited the territory from implementing its own bankruptcy laws. As economists Heriberto Martínez-Otero and Ian Seda-Irizarry noted, this left Puerto Rico with few options to maneuver around its massive public debt.[122]

In response to Puerto Rico's debt crisis, in June 2016 Congress passed and President Barack Obama signed the Puerto Rico Oversight, Management, and Economic Stability Act, which empowered the president to appoint a seven-member financial oversight and management board, based in New York City, to oversee the commonwealth's budget, restructure its debt, and impose whatever austerity measures were necessary to eliminate the debt, including the selling of public assets. Highly unpopular, the act further undermined the fiscal autonomy of the Puerto Rican commonwealth. Members of Congress said they saw no alternative to what it called a "catastrophic" debt burden; others viewed the intervention as an example of what Naomi Klein has called the "shock doctrine," the exploitation of national crises to push through controversial policies that benefit only a few. Yarimar Bonilla and Marisol LeBrón have argued that the board had "no vision for the island's future other than restoring its ability to continue borrowing and generating profit for investors."[123]

By May 2017, Puerto Rico had over $74 billion in bond debt (including those from municipalities, special authorities, and public corporations) and $49 billion in unfunded pension liabilities.[124] The island's electrical grid was old and needed repair, its infrastructure neglected, and the island highly polluted from industrial and military contaminants. Unemployment rates were more than two times the national average; the median household income was a third of the U.S. median income. The cost of living was significantly higher than in the fifty states, and the standard of living was spiraling downward for the middle and working classes. For many, internal migration to one of the fifty states was the only alternative.

Puerto Rico, on the eve of Hurricanes Irma and Maria, then, was already struggling. As President Trump had tweeted, the territory was "in deep trouble," facing a "massive debt" to "Wall Street and the banks which, sadly, [had to] be dealt with."[125] The hurricanes augmented a century of policy-driven vulnerabilities. As scholar Luis E. Rodríguez Rivera noted, Hurricane Maria "detonated" preexisting disasters and exacerbated the ongoing violation of Puerto Ricans' civil, political, and human rights. Lawyer Juan Pablo Bohoslavsky reminded observers that "even before Hurricane Maria struck, Puerto Rico's human rights were being massively undermined by the economic and financial crisis and austerity policies, affecting the rights to health, food, education, housing, water and social security."[126]

Like Puerto Rico, the challenges faced by the Virgin Islands were due, in part, to its peculiar territorial status and the debt its government took on to compensate for the lack of equal access to the entitlements of U.S. citizenship. Its dependence on tourism also made the territory economically vulnerable to natural disasters.

Although the islands were no longer significant agricultural producers, the United States acquired the Virgin Islands in 1917, purchasing the "Danish West Indies" from the Kingdom of Denmark for $25 million in gold to prevent them from becoming German outposts close to U.S. territory, should Germany have annexed Denmark during the First World War, and to use as naval bases that would help protect the shipping lanes in the Caribbean and Atlantic. (A fourth island, Water Island, located in St. Thomas's Charlotte Amalie harbor, was acquired in 1996.)[127]

The U.S. Navy administered the USVI until 1931, when the U.S. Department of the Interior—and a series of federally appointed civil administrators—assumed control of the territory's governance. The islands' residents were "American nationals" with limited rights until 1932, when an act of Congress finally granted them U.S. citizenship. Like Puerto Ricans, this was a limited form of citizenship that conferred the right on islanders to travel freely to the continental United States and to travel abroad with a U.S. passport but also brought the responsibility of serving in the U.S. military in times of war without the right to elect the president that could send them to war. The Organic Act of 1936 gave USVI residents greater political self-determination, but they were unable to elect their own governor until 1970. Their elected delegate to the U.S. House of Representatives, similar to Puerto Rico's resident commissioner, can vote only in the committees on which he or she serves.

Under U.S. control, St. Thomas, and its capital city of Charlotte Amalie, emerged as the center of government, finance, trade, and commerce. St. John became a center for tourism because the Virgin Islands National Park, operated by the U.S. National Park Service, occupies two-thirds of the island. St. Croix has traditionally been the center of manufacturing and agriculture. When the U.S.-appointed civil administrators failed to revitalize agriculture, manufacturing and tourism became the cornerstones of the economy. Assembly plants, two rum distilleries, and the Hovensa oil refinery provided jobs and revenue for decades, as did the tourism sector, which expanded after the U.S. economic embargo on Cuba redirected tourism away from Cuba and to the American territories in the Caribbean. Since 2000, tourism has been the principal driver of the USVI's economy, producing 60–80 percent of its GDP. More than 1.2 million cruise passengers and 400,000 airline passengers visit the islands each year.[128] Unfortunately, tourism makes the territory especially vulnerable to environmental disruptions, as well as to any economic downturn that affects people's ability to travel.

Even in the best of times, when manufacturing jobs provided more steady incomes, Virgin Islanders struggled economically. Food supplies and consumer goods, including the molasses needed for rum production, had to be imported, and, as in Puerto Rico, residents paid much higher prices for goods than Americans did in the fifty states. They also paid much higher prices for public utilities. The outdated and inefficient island power system has been fueled by diesel oil rather than the more obvious solar, wind, and geothermal energy—the renewable energy that Caribbean islands have in abundance. This means that residents have paid about three times the average cost of such services.[129] By 2009, the median household income for Virgin Islanders had fallen to 75 percent of the median income in the fifty states.[130] Like Puerto Ricans, residents qualified for some public welfare programs such as nutrition and housing assistance but at much lower rates than U.S. citizens in the states, and assistance was subject to statutory caps. Similar to Puerto Rico, the territorial government indebted itself to partially finance its operations. By 2016, its public debt was

72 percent of the GDP.[131] While the USVI had not defaulted on its loans, the territory had a public debt of $2 billion and a net pension liability of $4 billion. In 2015, the Mapp administration imposed austerity measures, slashing public spending and raising taxes on certain goods.

To address these economic challenges, the USVI's Economic Development Authority has promoted investment and manufacturing in the USVI by stressing that the territory "has all the benefits of being part of the United States." For manufacturers, these include the ability to use the "Made in USA" labeling that is important to many American consumers; duty-free imports; duty-free and quota-free exporting of USVI-made goods into the fifty states; a lack of state or territory tax; and, unlike Puerto Rico, exemption from the shipping regulations of the 1920 Jones Act. A select group of investors can qualify for 90–100 percent exemption from corporate, personal, excise, and property taxes.[132] Still, these policies have failed to generate the necessary income. On the eve of Hurricanes Irma and Maria, the USVI was struggling economically and could ill afford the economic impacts of these environmental disasters.

The Federal Reserve Bank has noted that after natural disasters, economic disruptions tend to be severe for one or two months but then dissipate as economic activity is boosted by insurance payouts, federal aid, and the jobs created by cleanup and reconstruction efforts. Conditions in the territories made Puerto Rico and the USVI an exception. As a result of Maria, Puerto Rico and the USVI suffered a 4 to 8 percent job loss, respectively, with the leisure/hospitality, trade, education, and health sectors most affected. Though job loss was not as severe as in New Orleans after Hurricane Katrina (29.7 percent), it was substantial nonetheless for these territories that had already lost thousands of jobs over the previous decade due to the closing of plants and to government austerity measures. The job loss figures also did not factor in the decrease in income that came as a result of decreased work hours, especially if residents worked in the informal economy.

There was also considerable deferred infrastructure maintenance in both territories, which contributed to the loss of jobs and the impacts on health and quality of life. Puerto Rico, for example, experienced the most severe power outage in U.S. history, and this affected the productivity of businesses, the reliability of government services, and the quality of life in homes and schools. USVI delegate Stacey Plaskett noted that the Virgin Islands may have had considerably fewer casualties than Puerto Rico, but many of the islands' challenges were related to their old and vulnerable buildings and infrastructure: "I think that's part of the reason that our hospitals sustained the amount of damage that they did," said Plaskett. "Yes, there was a Category 5 hurricane, but also we've never had the funding that we needed to do maintenance on the roof the way we should. We lost nine schools because there hasn't been a new school built in the Virgin Islands in over twenty-five years. A lot of those schools were already in disrepair and had issues related to structural integrity."[133] After back-to-back hurricanes, rebuilding required investments in state-of-the-art sustainable construction—that is, if the territories hoped to minimize the impacts of future storms that would come their way. Without federal government assistance, however, their debt precluded these adaptations.

Given these economic realities, could the Trump administration and Congress have done more for these American territories in the immediate aftermath of the storm? The Trump administration's response to the devastation in the USVI and Puerto Rico was anemic in part because of competing environmental emergencies in 2017, but the lack of political will also had a lot to do with it. (This lack of political will was also demonstrated in 1989 and 1998, when the Bush and Clinton administrations, respectively, delayed humanitarian responses to the islands after Hurricanes Hugo and Georges.)[134] Puerto Ricans and Virgin Islanders blamed this disinterest on their lack of political influence on the federal election process. This lack of influence means that presidential administrations can easily bracket or outright dismiss the concerns of territorial citizens. Some critics also attributed the indifference to racism. As Congressman Al Green noted, the president's criticisms of the territories were often racially charged: "You don't hear that kind of dog whistle, of people not wanting to pull themselves up by their bootstraps, when the people are Anglos. That's something reserved for people of color."[135]

The Trump administration, in turn, blamed the inefficiencies on the mismanagement and corruption in the territories, at least in the case of Puerto Rico. He rarely mentioned the USVI in any discussion of disaster relief. Congress passed supplemental aid packages to assist the territories, but the Trump administration withheld $16 billion allocated for Puerto Rico on the grounds that Puerto Ricans had received—and mismanaged—over $92 billion in aid: "more money than has ever been gotten for a hurricane before," said the president, but Puerto Ricans had "squandered away or wasted much of it." The president was misinformed. By April 2019, when the president tweeted his frustration on social media, Puerto Rico had received only $11.2 billion out of the $41 billion allocated by Congress for disaster relief.[136] Some of the funds were eventually released, but not until January 2020, when a series of earthquakes caused additional hardships on the island. Even then, in order to access the funds, Puerto Rican officials had to navigate a series of bureaucratic restrictions that disaster-affected states like Florida and Texas did not have to navigate.[137] Only in February 2021, three and a half years after Hurricane Maria, did the newly elected Biden administration announce that the government would release $1.3 billion in aid and lift Trump-administration restrictions on another $4.9 billion in order to help the U.S. territory become more resilient to future storms.

By virtue of population size, Puerto Rico received the lion's share of whatever assistance flowed to the territories during the Trump administration, but the cumulative resources were far fewer than those made available to other disaster victims in the fifty states. According to FEMA records, 92 percent of Puerto Ricans' homes (1,138,843) were damaged by Hurricane Maria, and by May 2018, 98 percent (1,118,862) had applied for assistance under FEMA's Individuals and Households Program; yet FEMA denied 40 percent of these applications, in many cases because residents could not meet FEMA's strict homeownership regulations.[138] After Hurricanes Katrina and Sandy, the federal government set up the Disaster Housing Assistance Program to help survivors pay rent, put down security deposits, or pay for utilities while they waited for the completion of home repairs, but the program was not offered to the territories. Puerto Ricans and Virgin Islanders did temporarily qualify for FEMA's Transitional Sheltering Assistance (TSA) program, which paid for motel rooms. Thousands of families (7,000 in Puerto Rico alone) received temporary shelter through TSA, but the program was terminated after nine months, forcing many of TSA's beneficiaries to scramble for alternative housing.

Puerto Rican plaintiffs filed a class action suit to try to block TSA's premature termination, noting that "hundreds of Hurricane Harvey victims remain[ed] sheltered in hotels and motels," so why were Puerto Ricans singled out? Instead, FEMA had offered Puerto Ricans return airfares for all those who wished to return home, but "the offer [was] illusory," said the plaintiffs, "as it provid[ed] a 'ticket to nowhere'; there [were] no homes to which these families [could] return." Despite a temporary injunction, the class action suit ultimately failed, and the TSA program ended in September 2018. Recognizing that longer-term assistance was necessary, U.S. congressman Adriano Espaillat, Senator Elizabeth Warren, and several of their colleagues introduced the Housing Victims of Major Disasters Act of 2018, but the bill failed to pass the Republican-controlled House and Senate. Similarly, the Puerto Rico and Virgin Islands Equitable Rebuild Act, introduced by Senator Bernie Sanders and USVI delegate Stacey Plaskett, also died.[139]

Territorial status limited other forms of assistance to the disaster victims. Medicaid funding in both territories was capped. The school meals program and the Women, Infants, and Children Program operated in Puerto Rico and the USVI, but the Supplemental Nutrition Assistance Program (SNAP), available to low-income households in the fifty states and the USVI, was not available to Puerto Rico. Instead, Puerto Rico was eligible for a "capped bloc grant" known as the Commonwealth Nutrition Assistance Program (NAP), but the criteria for qualification was much more stringent than in the fifty states and the benefits more limited. Unlike SNAP funding, which expands to accommodate need, NAP operated on a fixed level of annual funding for food assistance regardless of need. Once that fixed amount of money was used up, it was not supplemented by Congress. In the months after the hurricane, when need was greatest, NAP offered a weak safety net for low-income families who lost homes and livelihoods. Roughly 45 percent of Puerto Ricans lived below the poverty line before the hurricane compared with the national average of 14 percent, but they qualified for far less in assistance than their fellow Americans in the fifty states. Navigating the cumbersome bureaucracy to receive assistance was also difficult. A December 2017 assessment report published by Refugees International noted that "Maria survivors [encountered] enormous challenges navigating [FEMA's] bureaucratic and opaque assistance process and lack[ed] sufficient information on whether, when, and how they [would] be assisted."[140]

A year after the storm, a *Washington Post*/Kaiser Family Foundation survey of Puerto Rico residents reported that eight in ten people residing on the island had experienced damages to their homes, loss of employment, or health setbacks.[141] As happens in most environmental crises, women, children, and the elderly had been disproportionately affected. Prior to Hurricane Maria, Puerto Rico had the highest level of child poverty in the nation, at 57 percent, but the economic precarity of women, children, and the elderly worsened after the storm. A study commissioned by Puerto Rico's Youth Development Institute found that Puerto Rico's poorest children were now living in conditions "not seen in the United States in over half a century." Parents' loss of employment and diminished work hours impacted low-income families, especially those headed by women.[142]

Children in Puerto Rico and the USVI were also affected by the loss of schools and the migration of classmates and teachers. Because school buildings were heavily damaged in the USVI, authorities packed students into the few schools that remained functional. Teachers reported high rates of depression, anxiety, and other manifestations of PTSD.[143]

In Puerto Rico, the government permanently closed 300 schools to meet the debt restructuring requirements of the federal government, but closure was also driven by demographic changes: as families departed for the mainland, classroom occupancy rates dropped. Students who remained on the island often found themselves traveling longer distances to attend school, now overcrowded because of consolidation. During the 2017–18 academic year, Puerto Rican students missed, on average, seventy-eight days of school. At colleges and universities across the island, classrooms and laboratories were so damaged that educational institutions across the fifty states offered to temporarily accommodate Puerto Rican students either tuition-free or at in-state tuition rates.[144]

Internal Migration as Americans

When imploring Congress to pass a package commensurate with the scale of the disaster, Governor Rosselló occasionally played on American fears of a mass migration. There would be a "massive exodus to the United States" if the island was unable to recover, he warned, and since Puerto Ricans were U.S. citizens they could migrate to the States without any legal barriers.[145] Rosselló was correct—the migration of Puerto Ricans (and Virgin Islanders) was an internal migration from one American geographic space to another. But like other internal migrants around the world, those who migrated from the Caribbean islands faced social and economic challenges. A shared citizenship did not make their integration in the "mainland" automatic.

A month after the hurricane, researchers at the Center for Puerto Rican Studies at Hunter College projected that between 114,000 and 213,000 Puerto Ricans would leave the island annually in the wake of Hurricane Maria and that by 2019 Puerto Rico would lose over 470,000 residents, or 14 percent of its population.[146] But as the U.S. Census Bureau later noted, it has been difficult to assess how many people have actually migrated from the territories in the aftermath of Hurricane Maria and whether their migration was temporary or permanent. The Census Bureau initially reported an increase of 133,451 Puerto Ricans in the fifty states between 2017 and 2018 and an increase of 24,689 from other "U.S. Island Areas."[147] For Puerto Rico, this constituted a loss of close to 4 percent of its population in just twelve months—a figure that was more than double the annual rate over the past seven years.[148] Researchers at the Census Bureau later released more nuanced estimates. Drawing on monthly airline passenger traffic data from the Bureau of Transportation Statistics, the bureau determined that by 2020 Puerto Rico had experienced an annual net out-migration of 123,399 people to other parts of the United States. "Puerto Rico has seen a steady decline in population over the last decade," said one demographer. "[But] hurricane Maria . . . further impacted that loss, both before and during the recovery period."[149]

Other population studies have examined complementary sources, including school enrollment data, cell phone usage, and even social media sites.[150] Drawing on school enrollment figures, for example, the Center for Puerto Rican Studies reported that 135,000 Puerto Ricans relocated to the continental United States in the first six months after the hurricane.[151] Using cell phone tracking data, another study estimated that over 400,000 Puerto Ricans left the island between October 2017 and February 2018. According to this study, most of the migrants settled in areas that already had large Puerto Rican populations: Florida (43 percent), followed by New York (9 percent), Texas (7 percent), and Pennsylvania (6 percent). Within Florida, the metropolitan areas that received the largest populations were Orlando (22 percent in Orange County), Kissimmee (15 percent in Osceola County), and Miami (10 percent in Miami-Dade County).[152] The arrivals were significant enough that Florida governor Rick Scott declared a state of emergency to release funds for housing, education, crisis counseling, and employment assistance.[153]

The Puerto Rican migrants who settled in Florida joined over 1.2 million compatriots who had settled in the state by the eve of Hurricane Maria, accounting for one of every five Latinos in the state.[154] By 2015, Orlando's Puerto Rican population was second only to the more established Puerto Rican population of the New York metropolitan area, whose roots dated back to the nineteenth century. The new Florida residents were lured by the jobs at Cape Canaveral and other military facilities, as well as Walt Disney World and other sectors of the tourism economy.

Drawing on data from a social networking site, yet another study estimated a 17 percent increase in the number of Puerto Rican migrants present in the continental United States from October 2017 to January 2018 (roughly corresponding to 185,200 people). This study also confirmed that the states with the largest population increases were those that already had large Puerto Rican populations: Florida, New York, Pennsylvania, Connecticut, and Massachusetts. The researchers found a noticeable increase in the number of migrants in the fifteen-to-twenty-nine age bracket, which was understandable given the unemployment on the island post-hurricane. While there has been some return migration to Puerto Rico, out-migration has remained high, especially from the most economically vulnerable areas of the island.[155]

By 2018, Puerto Rico's population stood at 3.2 million—its lowest figure since 1979.[156] The USVI had lost roughly 1.5 percent of its population between 2010 and 2017 and decreased by an additional 1 percent by 2019.[157] In an already small population, even one percentage point was keenly felt. Temporary relocation or permanent migration out of the two American territories was made possible by the residents' U.S. citizenship, which exempted travelers from visas, passports, and other restrictions. It is likely that many more would have left in the immediate weeks after Hurricane Maria if not for the high cost of airline flights out of the territories: when Luis Muñoz Marín International Airport reopened, for example, the popular San Juan-to-Miami route cost as much as $2,000, with additional costs for travel from Miami to other cities. Irate customers, including musician Ricky Martin, tweeted snapshots of the inflated fares. Threats of a national boycott over price-gouging eventually forced airlines to cap their fares.[158]

As U.S. citizens from American territories, Puerto Ricans and Virgin Islanders were not immigrants, and they were intimately familiar with American institutions, which gave them certain advantages. But when moving to the States, they often encountered the same challenges immigrants did. They had to find housing and employment and enroll their children in school. They had to navigate impersonal administrative bureaucracies and racially discriminatory policies, and deal with neighbors and coworkers who considered them foreigners and were often hostile to their presence. Puerto Ricans faced the additional challenge of language; though Puerto Rico is a bilingual society, English fluency varies across the population. Like immigrants, Puerto Ricans and Virgin Islanders settled near compatriots, hoping to find community and a safety net that would facilitate their integration for however long they might remain in the States. Ironically, once they established residency they qualified for the same social services as other Americans. Had they qualified for these benefits before migrating, many might not have had to leave their home.

Many political observers predicted that island-born Puerto Ricans now living in the States would express their discontent with federal policies at the voting booth, especially in the swing state of Florida, where they now had the right to vote once they established proof of residency. Political operatives called the arrival of so many Puerto Ricans a potential "game changer," especially for Democrats if "they played it right."[159] The number of eligible Puerto Rican voters did increase 30 percent in Florida from 2016 to 2018; and fourteen of the eighteen counties with the largest Puerto Rican populations registered much higher voter registration rates than the state average. New registrations were more likely to register as Democrat and Independent than Republican. It is difficult to ascertain just yet whether the Puerto Rican expatriates will shape electoral outcomes in the years to come. Despite the many criticisms of Trump's handling of Hurricanes Irma and Maria, he won 30 percent of the Puerto Rican vote in Florida, allowing him to win Florida's twenty-nine electoral college votes in the highly contested 2020 election. Back on the island, over 52 percent of over 1.1 million voters called for statehood in the nonbinding plebiscite of November 2020.[160]

Yet More Environmental and Political Fault Lines

Unlike other disaster-affected areas in the United States, which recovered more rapidly, Puerto Rico and the USVI continued to struggle for years after Hurricanes Irma and Maria. FEMA funded only 190 long-term recovery projects in Puerto Rico out of the 9,000 requests and only 218 projects in the USVI out of its 1,500 requests.[161] One editorialist noted that the Trump administration's continued choice of language—the references to Puerto Rico as "that country," as well as the charges that Puerto Ricans "only take from the USA"—seemed a calculated move to underscore the point that Puerto Rico was not really part of the United States.[162] Throughout the remaining three years of his presidency, Trump regularly disputed the casualty rates, accusing Democrats of inflating figures to discredit his administration. More concerning, the president never set foot in Puerto Rico again and never visited the USVI at all. Through all the criticism, the president insisted that "Puerto Rico [had] been taken care of better by Donald Trump than by any living human being."[163]

Trump's charges that Puerto Rico had mismanaged the crisis had some truth, however, and the ensuing revelations distracted from the territories' very real complaints that they had been treated as second-class citizens. The Puerto Rico Electric and Power Authority, "once considered a crowning achievement of effective public management," became embroiled in scandal when it contracted Whitefish Energy, a small Montana-based company with limited experience in large-scale projects (but with ties to members of the Trump administration), to restore Puerto Rico's power grid in a deal the journalists from the news website Vox called "shady as hell." A congressional investigation revealed numerous irregularities:

PREPA did not pursue the open-bidding process required for federal reimbursement; Whitefish subcontracted the work to Florida utilities and used higher-cost charter jets rather than cargo ships to transport materials to the island; and it billed PREPA for utility line work at the base rate of $319 per hour (for sixteen-hour workdays, seven days a week) when workers were actually paid $63 per hour.[164] At a Senate hearing, members of the Rosselló administration defended the decision to work with Whitefish, arguing that it was the only company willing to enter into contract with debt-stricken Puerto Rico without a substantial down payment. The $300 million contract with Whitefish was subsequently canceled.

Other signs of corruption and mismanagement became apparent during the summer of 2019. The FBI arrested six individuals on thirty-two counts of corruption and fraud. Julia Keleher, Puerto Rico's secretary of education who closed hundreds of public schools in the wake of Hurricanes Irma and Maria, was one of the six, charged with funneling $13 million in fraudulent contracts through her department and "profit[ing] at the expense of Puerto Rico's children." Angela Ávila-Marrero, the executive director of the Puerto Rico Health Insurance Administration, was also accused of funneling $2.5 million through her department, even as she testified on Capitol Hill of a shortfall in Medicaid health coverage on the island. "Government officials are entrusted with performing their duties honestly and ethically," said one U.S. attorney. "The charged offenses are reprehensible, more so in light of Puerto Rico's fiscal crisis."[165]

It was a series of other revelations, however, that finally ended the Rosselló administration. A week following the arrests, Puerto Rico's Centro de Periodismo Investigativo (whose 2018 lawsuit had challenged the government's official casualty rates) published 889 pages of leaked text conversations between Rosselló and his top aides that revealed their contempt for political rivals, journalists, celebrities, FEMA officials, and even their own constituents. (In one conversation about forensic pathologists, Christian Sobrino Vega, Rosselló's chief economic development advisor, made light of the dead piling up in morgues around the island. "Don't we have some cadavers to feed our crows?" asked Vega.)[166] The chats also revealed how members of the Rosselló administration tried to manipulate public opinion polls and shared confidential government information to their personal advantage. Dubbed "#Rickyleaks" and "#Chatgate," the scandal drove hundreds of thousands of Puerto Ricans into the streets of San Juan for twelve consecutive days to demand Rosselló's resignation. Facing impeachment on five charges, including embezzlement, Rosselló resigned on July 24, 2019, on the eve of Constitution Day, Puerto Rico's national

Public protest against Governor Ricardo Rosselló, San Juan, summer 2019. (Photo by bgrocker/Shutterstock.com)

holiday. Since the secretary of state, the next in line, had also resigned as a result of the text scandal, Secretary of Justice Wanda Vázquez Garced became governor.

Not long after this political crisis, the new governor faced yet another set of environmental challenges. On January 11, 2020, an earthquake measuring 5.9 on the Richter scale—the ninth earthquake in one day—struck the island's southwestern coast. These earthquakes were among the 123 earthquakes of magnitude 3 or higher to rock the island in a month. (The earthquake of January 7, measuring 6.4, was the strongest in over a century.) The earthquakes triggered landslides and power outages; destroyed or forced the closure of homes, schools, businesses, hospitals, and government offices; and even left physical changes in the natural landscape that were visible from space.[167] Punta Ventana, a natural rock formation in Guayanilla that was a popular tourist destination, was destroyed. Over 5,000 people were left homeless.

Like hurricanes, these seismic events were not unusual in Puerto Rico and the USVI; the islands sit at the border of the Caribbean and North American tectonic plates, and the faults in this area are especially active.

(Geologists measured over 7,000 earthquakes from March 2019 to March 2020.) These seismic activities have consequences for all the countries in the Caribbean basin, but in the case of Puerto Rico, the earthquakes so close on the heels of two devastating hurricanes delayed recovery further. Residents were still dealing with insurance claims, the loss of jobs, and the rubble around them. After the 6.4 earthquake on January 7, Governor Vázquez declared a state of emergency to allow, yet again, the extraordinary measures required for rescue and relief. The USVI residents, in turn, braced themselves for the possibility of a tsunami but were spared such an event.[168] They couldn't help but wonder what other disasters—environmental, political, or economic—lay in store. The answer would come a few months later with the COVID-19 global pandemic, which would strike the island's most vulnerable populations especially hard.

As Puerto Rico's and the USVI's experiences with Hurricanes Irma and Maria make evident, environmental disasters are also policy-driven disasters. Hurricanes and earthquakes may be unavoidable, but governmental and institutional actors exacerbate their consequences by failing the people they are supposed to serve. FEMA's report assessing the 2017 hurricane season acknowledged many mistakes and urged Americans to establish a "culture of preparedness" at all levels of government "to prepare for the disasters that we will inevitably face." The hurricanes "showed that governments need to be better prepared with their own supplies," said the report, "to have pre-positioned contracts with enforcement mechanisms, and to be ready for the financial implications of a disaster."[169] FEMA offered practical advice, but, in an era of accelerated climate change, when back-to-back Category 4 or 5 hurricanes will likely become increasingly common, more creative and ambitious projects are required to help people survive, recover, and remain at home. Instead of investing billions of dollars in restoring Puerto Rico's power grid to its condition prior to the storms, for example, Congress should have mandated *improving* the power grid to make it more resilient to natural disasters. Transitioning to solar, geothermal, and offshore wind energy resources could have been part of those improvements. As Luis A. Avilés wrote, this should be the model for all of the American territories in the Atlantic and the Pacific.[170]

The social and economic impacts of Hurricanes Irma and Maria could have been minimized through sounder city and regional planning. It is yet unclear whether the territorial and federal governments will heed the lessons of recent history. Will the territories develop alternative energy sources; invest in culturally sensitive and resilient building and engineering practices that can better withstand environmental disruptions; invest in sustainable agriculture; and develop human capital through education, health care, and jobs for the twenty-first century? These programs require not only the political will of actors at multiple levels of government but also the cooperation of banking and commerce. They require the abandonment of neocolonial policies and practices that trap populations and block more creative alternatives. The future depends on it.

CONCLUSION

Moving Forward

Natural Disasters May Be Inevitable; Good U.S. Policy Is Not

> In very graphic ways, disasters serve as indices of the success or failure of a society to adapt, for whatever reasons, to certain features of its natural and socially constructed environment in a sustained fashion.
>
> —Anthony Oliver-Smith

Ioane Teitiota and his wife, both Kiribati nationals, moved to New Zealand in 2007. Three years later, facing removal for overstaying their visa, Teitiota petitioned the New Zealand government for refugee status for the family "on the basis of changes to his environment in Kiribati caused by sea-level-rise associated with climate change." Teitiota argued that his island, South Tarawa, would become uninhabitable in the next ten to fifteen years; in the meantime, the island was overpopulated because other Kiritbati nationals had moved there by the tens of thousands to escape environmental pressures on their own islands. This expanding population, said Teitiota, had increased social unrest and violence and had strained water supplies and other natural resources, making life increasingly difficult.

Denied refugee status, Teitiota appealed the decision through New Zealand's various tribunals and courts. The Immigration and Protection

Tribunal ruled that the claimant had failed to meet the legal criteria for refugee status established by the *Refugee Convention*.[1] The High Court called his arguments "novel and optimistic" but "unconvincing." The Court of Appeal ruled that while it had "every sympathy with the people of Kiribati," Teitiota's claim for refugee status was "fundamentally misconceived" and an attempt "to stand the [United Nations] Convention on its head." On July 20, 2015, the Supreme Court of New Zealand upheld the decisions of the lower courts, and Teitiota and his wife were subsequently placed in removal proceedings and forced to take their three New Zealand–born children with them. The Supreme Court did note, however, that the decision "did not mean that environmental degradation resulting from climate change or other natural disasters could never create a pathway into the Refugee Convention or protected person jurisdiction" but would require an "appropriate case."[2] Though the Supreme Court did not specify what the appropriate case might be, the Teitiota case opened the door to several other climate-related asylum cases in New Zealand. By May 2017, seventeen people from the Pacific—including five from Kiribati—had made refugee claims in New Zealand: thirteen were rejected, and four cases are still pending.

Climate-displaced persons like Teitiota, with diminishing economic opportunities and who are anticipating further erosion of their lands and livelihoods, have tried to relocate elsewhere—by the millions. Teitiota's case illustrates how difficult it has been for climate-driven migrants to receive recognition and opportunities for resettlement when they migrate to other countries. The UN Human Rights Committee supported New Zealand's decision to reject Teitiota's case, ruling that Teitiota's human rights had not been violated because he did not face persecution if he returned to Kiritbati. The committee also acknowledged that, despite the many environmental pressures, the Republic of Kiribati could—theoretically—take additional "affirmative measures" to protect and relocate its population. Teitiota did not receive the ruling he hoped for, but the committee's judgment was remarkable nonetheless in that it reminded the international community that "environmental degradation, climate change and unsustainable development constitute some of the most pressing and serious threats to the ability of present and future generations to enjoy the right to life"; states will be in violation of their human rights obligations if they return individuals to places where their lives are at risk.[3]

In 2017, New Zealand made the unprecedented announcement that it would reserve 100 humanitarian visas for Pacific Islanders displaced by climate change. It is still unclear whom the government will prioritize for admission from the hundreds of thousands of Pacific Islanders who are losing their livelihoods to rising sea levels and other manifestations of accelerated climate change. One hundred visas are a drop in the proverbial bucket—an inadequate response to great need—but New Zealand has set an important example regardless. Other nations will have to follow New Zealand's lead, expanding opportunities for regular and authorized migration to accommodate those both temporarily and permanently displaced from their homes because of natural disasters and the long-term impacts that come with accelerating climate change.

In the United States, interviews with unauthorized immigrants from Central America have revealed how extreme weather events such as hurricanes and earthquakes and slower-onset disasters such as drought and crop disease have caused food insecurity, destroyed livelihoods, and forced people to migrate internally and across international borders. Hemispheric policy makers have not sufficiently addressed the environmental drivers of migration in the region; instead, they have focused their attention on combating other drivers, such as the criminal and political violence in particular countries, or they have focused on the policing of borders to prevent more "migrant caravans" from reaching their countries. These efforts will have limited success unless they also tackle the root causes of climate-driven migration, and this requires a reassessment not only of energy, water, and agricultural policies but also of economic development and so-called debt relief programs that have hardened inequality.[4] When catastrophes strike, humanitarian aid packages are helpful, of course, but they don't fully address the underlying conditions that have made some populations vulnerable in the first place. Relief and recovery programs should do more than just return conditions to the status quo.

The urgency of addressing these issues cannot be ignored. Rising global temperatures are having catastrophic consequences for people around the world, increasing the frequency and ferocity of "climate shocks" such as rapid-onset hurricanes, but they are also creating the conditions for longer-lasting environmental degradation from drought and desertification. Changes to the natural world are bringing social, economic, and political changes to the regions affected and, eventually, demographic changes as people are forced to move.[5] The 2021 report of the Intergovernmental Panel on Climate Change stated that it was now "unequivocal" that human behavior has warmed the planet, causing "widespread and rapid changes in the atmosphere, ocean, cryosphere, and biosphere"; some of those changes may be irreversible for millennia.[6] The earth could warm more than 1.5 degrees Celsius (2.7 degrees Fahrenheit) over the next ten years, undermining the 2015 Paris Agreement goals to keep the worst future damage from climate change in check. The evidence was "irrefutable" and a "code red for humanity," said UN secretary-general António Guterres.[7]

Since the beginning of the twenty-first century, roughly three-quarters of those displaced by climate disasters have been situated in the countries of South and Southeast Asia and sub-Saharan Africa, but climate impacts are driving more and more people in and out of Latin America.[8] At present, most climate-driven migration around the world has been internal, occurring within countries, but climate change is also generating new cross-border migration patterns.[9] Geographic proximity, historical ties, language affinity, family and community networks, ease of transportation, and legal access all determine where people will move.[10] Because of well-established migration networks, the United States and Canada can expect to see climate-driven migration from other parts of the hemisphere, especially from Mexico, Central America, and the Caribbean. Studies have already noted that hurricanes increase migration to the United States from those countries that have a significant expatriate community living in the states.[11] One recent study of climate-driven migration forecast that, by 2050, 1.5 million migrants from Central America and Mexico alone will try to migrate to the United States each year because of drought and the food insecurity that accompanies this condition.[12]

Projections of the number of people expected to migrate worldwide in response to accelerating climate change are sobering, ranging from 100 million to 1 billion people by 2050.[13] Estimates vary because, as Kathleen Newland has noted, no allowance is made for adaptation to changes or for the ability of governments and a few other major actors to influence migration patterns. A concerted international response to climate change could have dramatic results: as a World Bank study noted, with the mitigation of climate change, migration could be reduced as much as 80 percent by 2050.[14]

It may be impossible to forecast with any certainty the number of people who will be displaced by climate change, but we do know that those forced to migrate internally and across borders will put pressures on their new locales, wherever they settle. They may compete with local populations for housing, employment, arable land, and other natural resources. As the study of history has shown, migrants often reshape geographic, social, and legal landscapes in their new settings.[15] Sometimes those pressures can result in conflict.

Not surprisingly, military and national security experts have focused considerable attention on tracking how climate change leads to sectarian violence, genocide, and war as populations compete for diminishing resources.[16] The United Nations reported that 40 percent of all intrastate conflicts over the past sixty years have been linked to the control and allocation of natural resources.[17] In the United States, the defense and intelligence communities have studied climate change from both a bureaucratic and a strategic perspective: climate change increases not only the likelihood that military installations around the world will need to be buttressed or relocated to deal with rising sea levels, for example,[18] but also the likelihood of using the armed forces for humanitarian missions or to resolve conflicts and maintain social order.[19] Pentagon officials have warned that climate change is a "strategic threat" and a "threat multiplier," increasing the probability of political destabilization and civil unrest, especially in the world's most vulnerable regions.[20] And as Christian Parenti has noted, climate change is also creating "new geographies of violence."[21] Anticipating conflict in a world of accelerating climate change, nations are responding with greater militarization, the surveillance of populations, and the repression of civil liberties and civil rights to protect and maintain control over resources, supply chains, and private property.

Drawing on the rhetoric of "threat," policy makers have subsumed climate policies into national security agendas. Instead of using this moment to seek reparative justice—addressing the needs and rights of those left most vulnerable to climate change, for example—governmental, military, and industry leaders have looked for ways to leverage whatever opportunities climate change may offer. "For the security/military-industrial complex," wrote Nick Buxton and Ben Hayes, this means that "climate change is just the latest in a long line of threats constructed in such a way as to consolidate . . . grip on power and public finance." For corporations, climate change offers opportunities for "an endless supply of resilience- and disaster-related services." "Whether it is written into a U.S. defense manual, a retail giant's corporate risk strategy, or a World Bank agriculture proposal," wrote Buxton and Hayes, "security now provides a ubiquitous framework for policies that seek to consolidate the interests of the powerful."[22]

As we saw in chapter 2, the land and water grabbing in the service of economic development and the targeting of environmental activists who protest such actions are often part of that agenda. Climate change may create opportunities for new markets and technologies, but it also increases the likelihood of a more violent world that reinforces inequalities and scarcity and represses any examination of how and why climate change accelerated in the first place. The real climate danger is, as Daniel Aldana Cohen has noted, that class, race, and gender inequalities will harden and lead to eco-apartheid: "Those brutal inequalities, and the bullets that maintain them—not the molecules of methane—are what will kill people."[23]

National security agendas categorize climate-driven migration as a disaster in and of itself: one that must be controlled, usually by blocking the movement of people. But we should reframe the discussion to recognize the opportunities for cooperation, solidarity, and community building.[24] The negative impacts associated with migration can be minimized through managed resettlement: anticipating where populations will be displaced, creating opportunities for temporary and permanent resettlement, and assisting the relocated—and their host communities—to adapt and work together through education, training, and equal access to services.

International cooperation is essential for a successful response to climate change and to climate-driven migration. As Kathleen Newland has noted, policy responses need to proceed on multiple tracks at once: "The necessary short-term response to sudden-onset disasters is humanitarian assistance; in the medium term, global efforts should focus on building resilience through adaptation. In the long term, purposeful mitigation of global warming may bend back the curve of climate change."[25] But industry, banking, and governmental actors also need to recognize the roles they have played in impoverishing and displacing particular populations around the globe, leaving them on the front lines of a climate crisis they did not create. Helping populations become "resilient" should mean more than just returning to business as usual. As the Cochabamba Declaration acknowledged, local communities and citizens must be respected as equal partners with governments in the protection and regulation of natural resources.[26] At the local level, anxieties about climate change can be harnessed for climate justice, but worldwide we run the risk that just the opposite will happen—that eco-fascist and hyper-nationalist discourses will lead to the hoarding of resources, the scapegoating of immigrants and other marginalized populations, and the denial of legal and human rights.[27]

This book discussed the particular challenges that countries of the Western Hemisphere face when addressing accelerated climate change. Montserrat, the U.S. Virgin Islands, and Puerto Rico represent the geographic vulnerability of small island states in the Caribbean that experience hurricanes, volcanic eruptions, earthquakes, and rising sea levels; but the case studies also illustrate how their political vulnerability as territorial outposts of larger powers limits the resources available to them, especially during times of crisis. Though they are citizens of these larger powers by virtue of territorial birthright, the rights and privileges that should accompany citizenship are called into question during disasters; they have not qualified for the same assistance, benefits, and protections as "stateside" citizens. It is not surprising that in the wake of disasters, residents of these and other territories have called for a redefinition of their political status.

The citizenship of territorial residents—unequal though it might be—does offer one critical advantage in an era of accelerated climate change, however: in the event that their homelands become inhospitable and uninhabitable one day, citizenship offers them the automatic right of abode in Great Britain or the United States—a right that sovereign small island states do not enjoy. People forced to migrate from sovereign states are dependent on the political will of nations willing to take them in. In April 2021, for example, a volcanic eruption on the Caribbean island of St. Vincent forced the evacuation of 20,000 people, or one-fifth of the population of the Grenadines. In this case, the evacuees did not have the right of abode in any country. They depended on the humanitarian goodwill of St. Lucia, Barbados, Grenada, and Antigua to accommodate them. It is still unclear whether this disaster will lead to the permanent loss of this population and, if so, where the evacuees will be allowed to permanently settle.[28]

The Central American case study in this book highlighted the difficulties of adapting to climate change when countries experience back-to-back political and environmental disasters that make full recovery and resilience difficult, if not impossible. Nicaragua and Honduras (and El Salvador and Guatemala) were called "post-conflict societies" because democratic elections and accords signed in the 1990s created possibilities for peace, but in the decades that followed, other upheavals came at a rapid clip, making economic growth and political stability elusive. On the environmental front, the isthmus routinely faced hurricanes on two coasts, earthquakes, volcanic eruptions, drought, and crop disease that affected agricultural production, landownership, and control over natural resources, as well as access to safe housing, water, and health care. On the political front, these countries experienced upheaval caused by a variety of state and non-state actors. Together with the high levels of indebtedness, these conditions destabilized economies, made it difficult for families and local communities to thrive, and invigorated a large-scale internal and cross-border migration through Central and North America. These migrants have found their entry blocked. Their experiences navigating immigration bureaucracies in the region are a harbinger of challenges to come.

As climate change accelerates, the United States and other countries will have to follow New Zealand's lead and work together to create opportunities for resettlement for those who have permanently lost their homes and livelihoods due to irreversible changes to which they cannot adapt in their homelands. In the United States, legal accommodations like Temporary Protected Status and Deferred Enforced Departure offer safe haven, but they were designed to be temporary responses; they are not an adequate or realistic response to many challenges that climate change will present. However, they are not the only administrative or legislative options that Americans can draw on within existing U.S. law to accommodate those temporarily or permanently displaced by environmental upheaval. Humanitarian "parole," for example, is one possible recourse available to policy makers. The Immigration and Nationality Act grants the executive branch the discretionary authority to allow people to enter the United States, on a case-by-case basis, if doing so is considered to be in the national interest. Since 2002, humanitarian parole requests have been handled by the U.S. Citizenship and Immigration Services and are granted for one year (renewable on a case-by-case basis).[29] Like TPS, parole has never placed individuals on a path to permanent residency or citizenship, nor has it made them eligible for social welfare benefits, but it has granted them the right to work during the duration of parole, allowing them to earn a living and send remittances back home. In some cases, conditions in the home country make it impossible for humanitarian parolees to return. In such cases, congressional intervention, through the passage of so-called adjustment acts, can be used (as they have been in the past) to allow humanitarian parolees to adjust their status to legal permanent resident.[30]

For those permanently uprooted, the "unallocated reserve" in the annual refugee quota is another option available to policy makers. Since 1980, the annual refugee quota has usually earmarked a small number of visas (2,000 on average) for discretionary uses. The U.S. Refugee Admissions Program has used the unallocated reserve to accommodate additional refugees from particular regions of the world when the regional sub-quotas were insufficient to respond to the extraordinary need. Those uprooted by disasters are in "extraordinary need" of safe haven, and the Refugee Admissions Program could use the unallocated reserve to accommodate them. This would likely require an amendment to the 1980 Refugee Act that expands the refugee definition to recognize climate-driven migration as a forced migration. In matters of asylum, the Justice Department would have to issue guidelines to assist the immigration courts in evaluating asylum cases that include environmental displacement.

In addition to these tracks within the immigration system, U.S. policy makers could reserve a percentage of the Diversity Visa Lottery (created by the 1990 Immigration Act) for those displaced by environmental disasters; or, more significantly, Congress could amend immigration law to create a designated path for permanent residency alongside the tracks for high-skilled labor and family reunification. Any one of these options would be a first step in recognizing our humanitarian obligations to those on the front lines of the climate change crisis.

In September 2019, acknowledging a "shared responsibility," Senator Edward Markey and Congresswoman Nydia Velázquez introduced the first bills of the twenty-first century to accommodate climate-displaced persons (S. 2565 and H.R. 4732), which authorize the president to set the admissions level for climate-displaced persons for each fiscal year at the same time that the president determines the refugee quota. The architects of the complementary bills imagined a minimum of 50,000 visas for climate-displaced persons each year who would be eligible for the same resettlement assistance as refugees. Introduced during the Trump administration, the bill had no chance of passing in the 116th Congress, but the election of Joseph R. Biden created new opportunities for addressing climate change and climate-driven migration. Shortly after taking office, Biden issued an executive order authorizing a report on climate change and its impact on migration and internal displacement. This was the first time a president publicly acknowledged the climate-migration nexus.[31] At the time of this writing, it was yet unclear what policies might result from this study. Although the current political mood worldwide has led to a barring of immigrants in the interest of national security, countries—especially the more economically developed—must do more to honor their international commitments to provide refuge and safeguard human rights.

The populations of developing countries are on the front lines of climate change's impacts, but even citizens in wealthy nations like the United States are not insulated from the problems their government and industry leaders and their consumerism have created. In the United States, the state of Louisiana currently has the fastest rate of land loss in the country, losing twenty-five square miles per year due not only to sinking land mass and rising sea levels but also to the environmental impacts of logging and oil companies. These developments have threatened state agriculture with saltwater intrusion and have called into question the continued viability of New Orleans as one of the country's major ports.[32] Trying to stall these rapid changes, in 2007 the state began implementing a fifty-year master plan to try to protect its remaining coastline through the building of levees, barrier island restoration, and the salvaging of the eroding marshland.[33] Other coastal states, including Maryland, Virginia, New York, and Florida, are experiencing their own environmental challenges. A July 2017 report from the Union of Concerned Scientists warned that by 2035, with moderate sea level rise, an estimated 170 coastal communities in the United States—half of them already socioeconomically vulnerable—will "reach or exceed the threshold for chronic inundation."[34]

These environmental challenges have resulted in internal migration and demographic change. Years before Puerto Ricans and Virgin Islanders migrated as a result of Hurricanes Irma and Maria, Hurricane Katrina displaced over a million Gulf Coast residents. The population of New Orleans fell by over half, and though some people eventually returned, by 2012 the population stood at only 76 percent of its pre-storm figures. Likewise, after "Superstorm Sandy," the New Jersey boroughs that suffered the greatest devastation also experienced the greatest declines in population.[35] In Louisiana, members of the Biloxi-Chitimacha-Choctaw tribe in Isle de Jean Charles received the dubious distinction of being labeled by the news media as the United States' first "climate refugees." Over 98 percent of Isle de Jean Charles has eroded since 1955, leaving "a sliver of land just a quarter-of-a-mile wide and less than two miles long that is slowly being swallowed by the Gulf of Mexico." A federal grant from the National Disaster Resilience Competition allowed twenty-five families to be resettled to a sugar farm forty miles inland—a test case with the eventual goal of moving the larger population out of harm's way.[36] In Shishmaref and Kivalina, Alaska, villagers will also be relocated because of melting permafrost, rising sea levels, and ever-intensifying storms, all of which have accelerated erosion.[37] During a 2015 visit to Alaska, on the eve of the Paris climate summit, President Barack Obama said, "What's happening in Alaska isn't just a preview to what will happen to the rest of us if we don't take action, it's our wake-up call. The alarm bells are ringing."[38]

Unlike developing countries, industrialized nations have more economic resources for climate adaptation projects and—when necessary—for relocations of their populations.[39] However, the costs associated with the mitigation of risk, rebuilding, and resettlement will continue to be significant—and perhaps unsustainable—in an era of accelerating climate change. Between 1980 and 2020, the costs associated with 273 weather- and climate-related disasters in the United States exceeded $1.79 trillion. Americans have witnessed an exponential increase in disaster-related costs with each passing decade.[40]

Natural disasters may be impossible to prevent, but the effects of climate change can be mitigated; and because much of the devastation that occurs in the wake of natural disasters is the result of shortsighted policies, their impacts can also be minimized with strategic planning, sustainable practices, and responsible, accountable, and transparent governance. Addressing climate change is a domestic challenge but also an international responsibility. In working together, nations can effectively respond to this challenge. But will policy makers heed their scientists and exercise the political will to make a difference?

NOTES

Introduction

1. The National Oceanic and Atmospheric Administration offers the clearest explanation of the difference between weather and climate: weather refers to short-term changes in the atmosphere, while climate describes the weather over a longer period of time in a specific area. "What's the Difference between Weather and Climate?," National Centers for Environmental Information, March 23, 2018, updated August 7, 2020, https://www.ncei.noaa.gov/news/weather-vs-climate.

2. The earth has experienced profound climatic changes before. According to the National Aeronautics and Space Administration, the paleoclimate record, combined with data from global models, reveals that the earth has experienced ice ages and intense warming periods in the past. However, the paleoclimate record also tells us that the current warming of the earth is occurring much more rapidly than in the past: "When global warming has happened at various times in the past two million years, it has taken the planet about 5,000 years to warm 5 degrees. The predicted rate of warming for the next century is at least 20 times faster. This rate of change is extremely unusual." "How Is Today's Warming Different from the Past?," NASA Earth Observatory, accessed December 6, 2017, https://earthobservatory.nasa.gov/Features/GlobalWarming/page3.php; Kurt M. Campbell, Jay Gulledge, J. R. McNeill, John Podesta, Peter Ogden, Leon Fuerth, R. James Woolsey, et al., *The Age of Consequences: The Foreign Policy and National Security Implications of Global Climate Change* (Washington, D.C.: Center for Strategic and International Studies and Center for a New American Security, November 2007), 7; J. R. McNeill, "Can History Help Us with Global Warming?," in Campbell et al., *Age of Consequences*, 23–24.

3. According to the U.S. Environmental Protection Agency, for every increase of two degrees Fahrenheit in the global average temperature, there will be a decrease of 5–15 percent in the yield of crops currently grown, an increase of 3–10 percent in the amount of rainfall during the heaviest precipitation times (which dramatically increases flooding risks), a decrease of 5–10 percent in the stream flow in some river basins, and an increase of 200–400 percent in the area burned by wildfires. "Climate Change: Basic Information," U.S. Environmental Protection Agency, February 23, 2016, https://www3.epa.gov/climatechange/basics/.

4. McNeill, "Can History Help Us with Global Warming?," 23.

5. McNeill, 23–24.

6. The National Climate Assessment, for example, reported a steady increase in the intensity, frequency, and duration of North Atlantic hurricanes (including Category 4 and 5 storms) since the early 1980s but acknowledged that the causes were still unclear. Higher ocean surface temperatures contribute to the development of tropical depressions and hurricane activity, but a number of natural and human-caused factors can and do increase surface temperatures. The compiled data and observable patterns may be suggestive of a correlation between human activity and a specific weather pattern but not entirely conclusive. Likewise, it is difficult to make projections about the future with 100 percent certainty. Scientists rely on mathematical modeling to offer projections about the future, but models offer probability. Members of the scientific community have generally been conservative in their estimates and often qualify their projections about climate change with words such as "likely," "very likely," and "with high confidence." Although a single extreme weather event may be difficult to attribute entirely to anthropogenic climate change, a review of past scholarship suggests that climate projections have been far too conservative. See National Climate Assessment, "Changes in Hurricanes," GlobalChange.gov, accessed June 30, 2016, http://nca2014.globalchange.gov/report/our-changing-climate/changes-hurricanes; and John Carey, "Storm Warnings: Extreme Weather Is a Product of Climate Change," *Scientific American*, June 28, 2011. Kurt Campbell and his team of collaborators noted that the effects of global warming have unfolded much more rapidly than initially projected. Campbell and his colleagues examined "the full range of what is *plausible*" under three different scenarios: "expected, severe, and catastrophic climate cases." Campbell et al., *Age of Consequences*, 5–6. Glenn Scherer, in turn, pointed out that the Intergovernmental Panel on Climate Change (IPCC) projected that fossil fuel and industrial emissions in 2010 would range from a best-case scenario of 7.7 billion tons of carbon to a worst-case scenario of 9.7 billion tons, but global emissions from fossil fuels alone came to total 9.1 billion tons, at the far extreme of IPCC projections, placing the earth "on track for a rise of between 6.3° and 13.3°F, with a high probability of an increase of 9.4°F by 2100." Glenn Scherer, "IPCC Predictions: Then versus Now," Climate Central, December 11, 2012, https://www.climatecentral.org/news/ipcc-predictions-then-versus-now-15340.

7. Campbell et al., *Age of Consequences*, 7. Rafael Reuveny counted thirty-eight cases of mass environmental migration in human history by 2005: land degradation played a role in twenty-seven of the thirty-eight cases; drought in nineteen; deforestation in seventeen; water scarcity in fifteen; floods in nine; storms in seven; and famine in five cases. See Rafael Reuveny, "Environmental Change, Migration and

Conflict: Theoretical Analysis and Empirical Explorations," cited in Campbell et al., *Age of Consequences*, 19. The Norwegian Refugee Council has a pessimistic assessment of future migration: "People's mobility largely depends on resources and networks, and climate change is likely to negatively affect people's resources, increase their vulnerability and thereby reduce their mobility." Norwegian Refugee Council, *Climate Changed: People Displaced*, April 15, 2009, 7–8, https://www.nrc.no/resources/reports/climate-changed-people-displaced/; Alassane Drabo and Lunguere Mously Mbaye, *Climate Change, Natural Disasters and Migration: An Empirical Analysis in Developing Countries* (Bonn, Ger.: Institute for the Study of Labor, August 2011), 8, http://ftp.iza.org/dp5927.pdf.

8. Cooper Martin, program director of the Sustainable Cities Institute at the National League of Cities, offered this useful distinction between sustainability and resilience in Cody Boteler, "Lessons from the Experts: Making a Resiliency Plan," *Smart Cities Dive*, September 14, 2017, https://www.smartcitiesdive.com/news/lessons-from-the-experts-making-a-resiliency-plan/504787/.

9. Clayton S. Manning, C. M. Krygsman, and M Speiser, *Mental Health and Our Changing Climate: Impacts, Implications, and Guidance* (Washington, D.C.: American Psychological Association, and ecoAmerica, 2017), https://ecoamerica.org/wp-content/uploads/2017/03/ea_apa_mental_health_report_web.pdf, cited in Yessenia Funes, "Report: Climate Change Impacts Mental Health, with People of Color Hit the Hardest," Colorlines (website), March 30, 2017, http://www.colorlines.com/articles/report-climate-change-impacts-mental-health-people-color-hit-hardest.

10. Testimony of David Waskow of Oxfam America in U.S. Senate, Subcommittee on International Development and Foreign Assistance, Economic Affairs, and International Environmental Protection, *Drought, Flooding and Refugees: Addressing the Impacts of Climate Change in the World's Most Vulnerable Nations*, 111th Cong., 1st sess., October 15, 2009 (Washington, D.C.: Government Printing Office, 2010), 17–18; "Top 10 of 2009—Issue #7: The World Is Talking about Climate Change and Migration," Migration Policy Institute, December 1, 2009, http://www.migrationpolicy.org/article/top-10-issues-2009-issue-7-world-talking-about-climate-change-and-migration.

11. Hannah Brock, "Climate Change: Drivers of Insecurity and the Global South," Oxford Research Group, June 1, 2012, 9, http://www.oxfordresearchgroup.org.uk/publications/briefing_papers_and_reports/climate_change_drivers_insecurity_and_global_south; Joanna Nagel, *Gender and Climate Change: Impacts, Science, and Policy* (New York: Taylor and Francis, 2016).

12. Anthony Lake, "Foreword," in UNICEF Office of Research, *The Challenges of Climate Change: Children on the Front Line* (Florence, Italy: UNICEF Office of Research, 2014), p. vi, https://www.unicef-irc.org/publications/716-the-challenges-of-climate-change-children-on-the-front-line.html; Marie-Claude Martin, "Preface," in UNICEF Office of Research, *Challenges of Climate Change*, v.

13. Joint Global Change Research Institute, Battelle Memorial Institute, Pacific Northwest Division, and National Intelligence Council, *Mexico, the Caribbean, and Central America: The Impact of Climate Change to 2030: A Commissioned Research Report*, December 2009, 13, http://globaltrends.thedialogue.org/publication/mexico-the-caribbean-and-central-america-the-impact-of-climate-change-to-2030/.

14. Nansen Initiative, *Agenda for the Protection of Cross-Border Displaced Persons in the Context of Disaster and Climate Change*, December 2015, 1:6, https://environmentalmigration.iom.int/agenda-protection-cross-border-displaced-persons-context-disasters-and-climate-change.

15. Rob Nixon, *Slow Violence and the Environmentalism of the Poor* (Cambridge, Mass.: Harvard University Press, 2011), 2–4.

16. The 1951 convention limited assistance to European refugees in the aftermath of the Second World War. The 1967 protocol removed these temporal and geographic restrictions. Using the masculine pronoun, the UN defined a refugee as any person who, owing to well-founded fear of being persecuted for reasons of race, religion, nationality, membership of a particular social group, or political opinion, is outside the country of his nationality and is unable or, owing to such fear, unwilling to avail himself of the protection of that country; or who, not having a nationality and being outside the country of his former habitual residence as a result of such events, is unable or, owing to such fear, unwilling to return to it. UN, *Convention and Protocol Relating to the Status of Refugees* (Geneva: UNHCR Communications and Public Information Service, 2010), 14, http://www.unhcr.org/en-us/3b66c2aa10. The United States did not sign the 1951 convention, but it did sign the 1967 protocol.

17. UNHCR, *Handbook on Procedures and Criteria for Determining Refugee Status under the 1951 Convention and the 1967 Protocol relating to the Status of Refugees* (Geneva: UNHCR, 1992), 16, http://www.unhcr.org/4d93528a9.pdf; "UNHCR, Refugee Protection, and International Migration," 5, UNHCR, January 2007, http://www.unhcr.org/4a24efoca2.pdf; Norwegian Refugee Council, *Climate Changed*, 19.

18. Barney Thompson, "Climate Change and Displacement," UNHCR, October 15, 2019, https://www.unhcr.org/en-us/news/stories/2019/10/5da5e18c4/climate-change-and-displacement.html. The Organization of African Unity stated that the term "refugee" should also apply to "every person who, owing to external aggression, occupation, foreign domination or events seriously disturbing the public order in either part or the whole of his country or origin or nationality, is compelled to leave his place of habitual residence in order to seek refuge in another place outside his country of origin or nationality." See OAU, "Convention Governing the Specific Aspects of Refugee Problems in Africa," UNHCR, accessed September 15, 2017, http://www.unhcr.org/en-us/about-us/background/45dc1a682/oau-convention-governing-specific-aspects-refugee-problems-africa-adopted.html. The Cartagena Declaration stated that "in addition to containing elements of the 1951 Convention . . . [the definition] includes among refugees, persons who have fled their country because their lives, safety or freedom have been threatened by generalized violence, foreign aggression, internal conflicts, massive violations of human rights or other circumstances which have seriously disturbed the public order." See Colloquium on the International Protection of Refugees in Central America, Mexico, and Panama, Cartagena de Indias, Colombia, "Cartagena Declaration on Refugees," UNHCR, November 22, 1984, http://www.unhcr.org/en-us/about-us/background/45dc19084/cartagena-declaration-refugees-adopted-colloquium-international-protection.html. The Organization of American States has also passed a series of resolutions offering additional guidance on how to respond to refugees, asylum seekers, stateless persons, and others in need of

temporary or permanent protection. In Europe, the European Union Council Directive (2004) has identified the minimum standards for the qualification and status of refugees or those who might need "subsidiary protection." European Union, "Council Directive 2004/83/EC," EUR-Lex, April 29, 2004, https://eur-lex.europa.eu/legal-content/EN/TXT/?uri=celex%3A32004L0083.

19. The origins of the terms "climate refugee" and "environmental refugee" can be traced back several decades. In the 1970s, Lester Brown of the Worldwatch Institute argued that environmental change would produce "environmental refugees." Essam El-Hinnawi also used the term "environmental refugee" in his 1985 United Nations Environment Programme paper. He defined environmental refugees as "those people who have been forced to leave their traditional habitat, temporarily or permanently, because of a marked environmental disruption (natural and/or triggered by people) that jeopardized their existence and/or seriously affected the quality of their life." Three years later, Jodi Jacobson for the Worldwatch Institute offered the first estimates of present and future "environmental refugees" and linked their numbers specifically to climate change. In 1991, Astri Suhrke and Annamaria Visentin distinguished between environmental migrants (who voluntarily choose to leave their country) and environmental refugees (who are driven away by sudden disasters). See Lester Brown et al., "Twenty-Two Dimensions of the Population Problem," Worldwatch Paper 5, 1976, https://eric.ed.gov/?id=ED128282; Essam El-Hinnawi, *Environmental Refugees* (Nairobi: United Nations Environment Programme, 1985); Jodi L. Jacobson, *Environmental Refugees: A Yardstick of Habitability* (Washington, D.C.: Worldwatch Institute, 1988); Astri Suhrke and Annamaria Visentin, "The Environmental Refugee: A New Approach," *Ecodecision*, no. 2 (September 1991): 73–74; Patricia L. Saunders, "Environmental Refugees: The Origins of a Construct," in *Political Ecology: Science, Myth and Power*, ed. Philip Anthony Stott and Sian Sullivan (London: Hodder Education Publishers, 2000), 218–46; and James Morrissey, "Rethinking the 'Debate on Environmental Refugees': From 'Maximalists and Minimalists' to 'Proponents and Critics,'" *Journal of Political Ecology* 19 (2012), https://journals.librarypublishing.arizona.edu/jpe/article/id/1844/.

20. Among the terms used are "environmental migrant," "environmental displacees," "environmentally displaced person," "environmentally motivated migrant," and "environmentally-induced migrants." The International Organization for Migration's (IOM) 2004 and 2011 editions of the *Glossary on Migration* did not include entries for "climate refugees," but they did list the terms "environmental migrants" and "environmentally displaced persons." See IOM, *Glossary on Migration* (Geneva: International Organization for Migration, 2004), http://www.iomvienna.at/sites/default/files/IML_1_EN.pdf; "Discussion Note: Migration and the Environment," IOM, November 1, 2007, 1–2, https://www.iom.int/jahia/webdav/shared/shared/mainsite/about_iom/en/council/94/MC_INF_288.pdf. The IOM has suggested the term "environmentally-induced migrants," which the agency defines as "persons or groups of persons who, for compelling reasons of sudden progressive changes in the environment as a result of climate change that adversely affect their lives or living conditions, are obliged to leave their habitual homes, or choose to do so, either temporarily or permanently, and who move either within their territory or abroad." See "Migration, Change, and the Environment: Definitional Issues," IOM, accessed October 2, 2017, https://www.iom.int/definitional-issues.

21. Anthony Oliver-Smith, "Disasters and Forced Migrations in the 21st Century," Social Science Research Council, June 11, 2006, http://understandingkatrina.ssrc.org/Oliver-Smith/; Kathleen Newland, *Climate Change and Migration Dynamics*, Migration Policy Institute, September 2001, 1, http://www.migrationpolicy.org/research/climate-change-and-migration-dynamics; Roger Zetter, *Protection in Crisis: Forced Migration and Protection in a Global Era*, Migration Policy Institute, March 2015, https://www.migrationpolicy.org/research/protection-crisis-forced-migration-and-protection-global-era.

22. At the 2005 UN World Summit, parties affirmed the principle of the "responsibility to protect": that is, state sovereignty carried with it the obligation to protect one's own people, but if the state was unwilling or unable to do so, the responsibility shifted to the international community, which was required to use diplomatic, humanitarian, or other means to protect the vulnerable. The "R2P principle" was a response to genocide, war crimes, ethnic cleansing, and crimes against humanity—not climate change. However, the same summit that articulated the R2P principle also acknowledged the need to address climate change through the reduction of greenhouse gases, conservation and water management, the implementation of sustainable development goals, and the mitigation of risk. UN Office on Genocide Prevention and the Responsibility to Protect, "Responsibility to Protect," United Nations website, accessed November 3, 2021, https://www.un.org/en/genocideprevention/about-responsibility-to-protect.shtml.

23. As Simon Behrman noted, "The word *refugee* is derived from the French verb *se réfugier*, which simply means to seek shelter from danger. . . . The principle of protection surely covers those whose lands are being flooded, whose crops are failing, whose water supply is being contaminated, and thus whose lives are becoming literally unlivable, and who are seeking a refuge elsewhere. And if it does, then why is the word 'refugee' resisted to such an extent when it comes to identifying their needs and their rights?" Simon Behrman, "What's in a Name? Climate Change and 'Refugees,'" Climate Refugees Mini-Symposium (Part II), University of East Anglia, *Blog: International Law @ UEA*, accessed August 6, 2017, http://sites.uea.ac.uk/law/research/international-law-blog/-/asset_publisher/bS26fAaA3cQa/content/mini-symposium-part-two-what-s-in-a-name-climate-change-and-refugees (page discontinued).

24. The IOM has also urged consulting the legal instruments on internal displacement, disaster management, and legal migration. Dina Ionesco, "Let's Talk about Climate Migrants, Not Climate Refugees," United Nations Sustainable Development Goals, accessed July 20, 2020, https://www.un.org/sustainabledevelopment/blog/2019/06/lets-talk-about-climate-migrants-not-climate-refugees/.

25. Carolina Fritz, "Climate Change and Migration: Sorting through Complex Issues without the Hype," *Migration Information Source*, March 4, 2010, http://www.migrationpolicy.org/article/climate-change-and-migration-sorting-through-complex-issues-without-hype; Rabab Fatima, Anita Jawadurovna, and Sabira Coelho, "Human Rights, Climate Change, Environmental Degradation and Migration: A New Paradigm," *Issue in Brief*, March 2014, 2–3, https://www.migrationpolicy.org/research/human-rights-climate-change-environmental-degradation-and-migration-new-paradigm.

26. Since 2009, Sheikh Hasina, the prime minister of the densely populated Bangladesh, a country especially hard hit by flooding, has been especially vocal in pressuring the UN General Assembly to create some type of legal framework for addressing climate-induced migration. Hasina, a winner of the "Champions of the Earth" award, the UN's top environmental prize, has also become a leader in implementing the UN's sustainable development goals. An estimated one in seven Bangladeshis are forecast to be displaced by climate change by midcentury. "Bangladesh Prime Minister Wins Top United Nations Environmental Prize for Policy Leadership," UN Environment Programme, September 14, 2015, https://www.unep.org/news-and-stories/story/bangladesh-prime-minister-wins-top-united-nations-environmental-prize-policy; Sheikh Hasina, "Bangladesh Not Waiting for the World to Save Us," *Huffington Post*, September 16, 2016, https://www.huffingtonpost.com/bangladesh-prime-minister-sheikh-hasina/bangladesh-not-waiting-for-the-world-to-save-us_b_8149554.html.

27. Social scientist Roger Zetter argues that perhaps as much as 95 percent of displaced populations remain in their own countries or in neighboring countries. Zetter, *Protection in Crisis*, 1.

28. Frank Laczko and Elizabeth Collett, "Assessing the Tsunami's Effects on Migration," *Migration Information Source*, April 1, 2005, http://www.migrationpolicy.org/article/assessing-tsunamis-effects-migration; International Federation of Red Cross and Red Crescent Societies, *World Disasters Report 2002: Focus on Reducing Risk* (Geneva, Switz.: International Federation of Red Cross and Red Crescent Societies, 2002), 89, http://www.ifrc.org/Global/Publications/disasters/WDR/32600-WDR2002.pdf.

29. The Internal Displacement Monitoring Centre's (IDMC) definition of internally displaced persons (IDPs) is drawn from the UN's 1998 Guiding Principles on Internally Displaced: "The criteria related to the 'forced' nature of displacement 'within internationally recognized borders' is clearly fundamental in determining whether the person is an IDP, but the Guiding Principles do not set other criteria by which to identity a person fleeing their 'home or place of habitual residence.' As such, we interpret IDPs to include not only citizens of the country in which displacement takes place, but also non-nationals such as migrants and asylums seekers." See IDMC, *2016 Global Report on Internal Displacement* (Geneva: Internal Displacement Monitoring Centre, 2016), 14, 82, https://www.internal-displacement.org/sites/default/files/inline-files/2016-global-report-internal-displacement-IDMC.pdf. In 2016 alone, 24.2 million people were displaced by disasters, most of them in South and East Asia, and an additional 17.2 million disaster-related displacements occurred in 2018, most in Asia and Africa. IDMC, *Global Report on Internal Displacement*, Internal Displacement Monitoring Centre, 2017, http://www.internal-displacement.org/global-report/grid2017/. Alina O'Keefe, in turn, wrote that disasters were responsible for 23.5 million new displacements in 2016, "with 97 percent caused by weather and climate-related events." Alina O'Keefe, "Disasters and Climate Change: Internally Displaced Persons and Refugees," Aid and International Development Forum, July 18, 2017, http://www.nonprofitpro.com/article/disasters-climate-change-internally-displaced-persons-refugees/; IDMC, *Global Report on Internal Displacement 2019* (Geneva: Internal Displacement Monitoring

Centre, 2019), https://www.internal-displacement.org/sites/default/files/publications/documents/2019-IDMC-GRID.pdf.

30. The Inter-Agency Standing Committee (IASC), a Geneva-based coordinating committee of international humanitarian organizations, which coordinates humanitarian assistance worldwide, affirmed that climate change could result in mass displacement but has not offered any numerical projections; instead, its reports stress that most displacement is likely to occur internally rather than cross-border. IASC, "Climate Change, Migration, and Displacement: Who Will Be Affected?," UNHCR, October 31, 2008, http://www.unhcr.org/en-us/protection/environment/4a1e4fb42/climate-change-migration-displacement-affected-working-paper-submitted.html; Jane McAdam and Julia Blocher, "Fact Check Q&A: As the Climate Changes, Are 750 Million Refugees Predicted to Move Away from Flooding?," *The Conversation*, August 3, 2016, http://theconversation.com/factcheck-qanda-as-the-climate-changes-are-750-million-refugees-predicted-to-move-away-from-flooding-63400.

31. The IPCC reported, "Many areas to which they flee are likely to have insufficient health and other support services to accommodate the new arrivals. Epidemics may sweep through refugee camps and settlements, spilling over into surrounding communities. In addition, resettlement often causes psychological and social strains, and this may affect the health and welfare of displaced populations." IPCC, *Climate Change: The IPCC 1990 and 1992 Assessments*, June 1992, 103, https://www.ipcc.ch/site/assets/uploads/2018/05/ipcc_90_92_assessments_far_full_report.pdf. Two decades after the IPCC's first report, the *Stern Review* also warned that long-term environmental deterioration would lead to worldwide competition for natural resources and would politically destabilize regions. Economist Nicholas Stern and his research team wrote this 662-page report for the UK Parliament. Nicholas Stern, *The Stern Review: The Economic Effects of Climate Change* (Cambridge: Cambridge University Press, 2006), 112, http://mudancasclimaticas.cptec.inpe.br/~rmclima/pdfs/destaques/sternreview_report_complete.pdf.

32. Historian Samuel Moyn wrote, "No one has attempted to write down a history of human duties. Even that phrase sounds strange. In particular, there is now a whole canon on the history of the internationalization of human rights since the middle of the 20th Century. But, to the best of my knowledge, there is not a single book on the history of duties.... Our age of rights, lacking a public language of duties, is a historical outlier. The consequences are significant." Samuel Moyn, "Rights vs. Duties: Reclaiming Civic Balance," *Boston Review*, May 16, 2017, http://bostonreview.net/books-ideas/samuel-moyn-rights-duties.

33. Amber Jamil, "Climate Refugees and International Law," *American Interest*, July 14, 2017, https://www.the-american-interest.com/2017/07/14/climate-refugees-international-law/. Climate change policy, then, was an "intergenerational foreign-aid program," especially for lesser-developed nations, where much of the world's population resides. "The Spirit of Kyoto: The Kyoto Protocol and Its Implications for U.S. Climate Change Policy," 5–6, in "The Spirit of Kyoto" file, Counsel's Office Records, Jason Wilson, 2013-0518-F, box 2, William J. Clinton Presidential Library (hereafter Clinton Presidential Library); Thomas C. Schelling, "The Cost of Combatting Global Warming," *Foreign Affairs* 76, no. 1 (November/December 1997): 8–14; World People's

Conference on Climate Change and the Rights of Mother Earth, "Universal Declaration of Rights of Mother Earth" (the "Cochabamba Declaration"), Cochabamba, Bolivia, April 24, 2010, https://pwccc.wordpress.com/2010/04/24/peoples-agreement/.

34. "UNHCR Viewpoint: 'Refugee' or 'Migrant'? Which Is Right?," UNHCR, July 11, 2016, http://www.unhcr.org/en-us/news/latest/2016/7/55dfoe556/unhcr-viewpoint-refugee-migrant-right.html. The Refugee Convention drew inspiration from the UN's 1948 Universal Declaration of Human Rights, which recognized the right of individuals to life, liberty, and security and the right "to seek and enjoy asylum" from persecution. "Universal Declaration of Human Rights," United Nations, accessed November 4, 2021, https://www.un.org/en/about-us/universal-declaration-of-human-rights; UNHCR, "Introductory Note," *Convention and Protocol Relating to the Status of Refugees*, 2, UNHCR, http://www.unhcr.org/en-us/3b66c2aa10. International law recognizes an individual's right to request asylum—and a state's sovereign right to grant it—but an individual has no internationally recognized right to *receive* asylum. The UN Conference on Territorial Asylum also failed to establish an international right to asylum. See Guy S. Goodwin-Gill, *The Refugee in International Law* (Oxford: Oxford University Press, 1983).

35. For an excellent discussion of the UNHCR's response to climate change, see chapter 2 of Nina Hall's *Displacement, Development, and Climate Change: International Organizations Moving beyond Their Mandates* (London: Routledge, 2016). According to Hall, the agency "did this in two ways: 1) the issue of environmental displacement was discussed as a protection issue, and 2) the agency developed an environmental division and environmental impact guidelines. These changes occurred in parallel to the expansion of its humanitarian operations under the leadership of a new High Commissioner, Sadako Ogata (1990–2000)." At a 2007 meeting of the executive committee of the National Security Council, António Guterres warned member states that desertification and sea level rise would displace millions. "The international community seems no more adept at dealing with those new causes than it is at preventing conflict and persecution," said Guterres. Cited in Hall, *Displacement, Development, and Climate Change*, 58. The following year, in an interview with *The Guardian*, Guterres once again warned that climate change was one of the main drivers of forced displacement, uprooting people from their traditional homes and livelihoods and triggering extreme poverty and conflict. Julian Borger, "Conflicts Fueled by Climate Change Causing New Refugee Crisis, Warns U.N.," *The Guardian*, June 16, 2008, https://www.theguardian.com/environment/2008/jun/17/climatechange.food. Despite his many subsequent public acknowledgments of climate change in interviews, speeches, and high-level meetings, Guterres was unable to convince the donor countries on which the agency is heavily reliant that new protections for forced migrants are warranted. As Hall pointed out, "UNHCR is particularly reliant on its donors to continue its operations and has no permanent funding, which means that long-term multi-year planning is difficult. Approximately 98 percent of the office's funding is from voluntary contributions." See Hall, *Displacement, Development, and Climate Change*, 67–72; and the Norwegian Refugee Council, *The Nansen Conference: Climate Change and Displacement in the 21st Century*, June 5–7, 2011, 5, http://www.unhcr.org/4ea969729.pdf. The 2011 Oslo conference was in honor of Fridtjof Nansen, the UN's first High

Commissioner for Refugees. At the Oslo conference, Guterres once again called for a new legal framework to assist those forced to cross borders because of climate change: "Even if they are not refugees, such people are entitled to our support and to have their voices heard and taken into account." António Guterres, "Nansen Conference on Climate Change and Displacement; Statement by António Guterres, United Nations High Commissioner for Refugees," UNHCR, June 6, 2011, https://www.unhcr.org/en-us/admin/hcspeeches/4def7ffb9/nansen-conference-climate-change-displacement-statement-antonio-guterres.html.

36. According to Hall, between 2005 and 2010 the UNHCR had an operational involvement in thirteen of fifty-eight natural disasters. "In all of these cases, UNHCR already had an operational presence in the country and thus there was a humanitarian imperative to assist." See Hall, *Displacement, Development, and Climate Change*, 72–73.

37. The UNHCR's *Guiding Principles on Internal Displacement* defined IDPs as "persons or groups of persons who have been forced or obliged to flee or to leave their homes or places of habitual residence, in particular as a result of, or in order to avoid, the effects of armed conflict, situations of generalized violence, violations of human rights *or natural or human-made disasters* [emphasis mine], and who have not crossed an internationally recognized state border." The *Guiding Principles* affirms that those displaced within their own countries are entitled to a wide range of rights and protections, including the "right to life, dignity, and security of person" and the right to humanitarian assistance and legal protection." UNHCR, *Guiding Principles on Internal Displacement*, 1998, 1, http://www.unhcr.org/en-us/protection/idps/43ce1cff2/guiding-principles-internal-displacement.html. States are not required to adopt these principles, nor are they binding once adopted. Royce Bernstein Murray and Sarah Petrin Williamson, "Migration as a Tool for Disaster Recovery: A Case Study on US Policy Options for Post-Earthquake Haiti," Working Paper 255, Center for Global Development, June 1, 2011, 4, https://www.cgdev.org/publication/migration-tool-disaster-recovery-case-study-us-policy-options-post-earthquake-haiti; Norwegian Refugee Council, *Climate Changed*, 10–11. While the principles are nonbinding, they have helped humanitarian and development agencies around the world establish operational procedures for assisting populations during times of crisis. The IASC drew on the *Guiding Principles* when it drafted its own framework for durable solutions, for example. IASC, "Framework for Durable Solutions for Internally Displaced Persons," UNHCR, accessed October 17, 2017, http://www.unhcr.org/en-us/protection/idps/50f94cd49/iasc-framework-durable-solutions-internally-displaced-persons-april-2010.html. The World Bank also drew on the *Guiding Principles* when it drafted its "Operational Manual 4.12" to regulate the conduct of officials working in resettlement and relocation efforts. The World Bank Operational Manual, "Operational Policy 4.12–Involuntary Resettlement," World Bank website, December 2001, http://web.worldbank.org/archive/website00527/WEB/OTHER/227908E4.HTM?OpenDocument. As Elizabeth Ferris noted, the World Bank's "Op 4.12" provides bank staff a detailed list of authorized activities. The emphasis is on preventing impoverishment rather than on upholding basic rights. Elizabeth Ferris, *Protection and Planned Relocations in the Context of Climate Change*, UNHCR, August 1, 2012,

https://www.brookings.edu/research/protection-and-planned-relocations-in-the-context-of-climate-change/. In 2009, the African Union produced the first legally binding contract, the Kampala Convention, that defines the obligations of the fifty-five member states to the internally displaced populations on the African continent. *African Union Convention for the Protection and Assistance of Internally Displaced Persons in Africa (Kampala Convention)*, accessed March 22, 2018, https://au.int/en/treaties/african-union-convention-protection-and-assistance-internally-displaced-persons-africa. See also Frank Biermann and Ingrid Boas, "Protecting Climate Refugees: The Case for a Global Protocol," *Environment* 50, no. 6 (December 2008/January 2009), 8–17, http://www.environmentmagazine.org/Archives/Back%20Issues/November-December%202008/Biermann-Boas-full.html; David Keane, "The Environmental Causes and Consequences of Migration: A Search for the Meaning of 'Environmental Refugees,'" *Georgetown International Environmental Law Review* 16, no. 2 (Winter 2004): 209–23; and Brock, "Climate Change," 8–9.

38. Norwegian Refugee Council, *Climate Changed*, 15–16; *Convention relating to the Status of Stateless Persons*, United Nations Treaty Collection website, September 28, 1954, https://treaties.un.org/pages/ViewDetailsII.aspx?src=TREATY&mtdsg_no=V-3&chapter=5&Temp=mtdsg2&clang=_en.

39. As Anthony Oliver-Smith noted in his 1992 study of the Andes, migration from rural to urban areas was already well underway when the Peruvian earthquake of 1970 struck, but the disaster accelerated this migration because it was in urban areas that rural peoples found disaster relief. Anthony Oliver-Smith, *The Martyred City: Death and Rebirth in the Andes* (Prospect Heights, Ill.: Waveland Press, 1992), cited in Anthony Oliver-Smith, "Disasters and Forced Migrations in the 21st Century," Social Science Research Council, June 11, 2006, http://understandingkatrina.ssrc.org/Oliver-Smith/.

40. The 1953 act defined a refugee as "any person in a country or area which is neither Communist nor Communist-dominated, who because of persecution, fear of persecution, *natural calamity* [emphasis mine] or military operations is out of his usual place of abode and unable to return thereto, who has not been firmly resettled, and who is in urgent need of assistance for the essentials of life or for transportation." Refugee Relief Act of 1953, Public Law 83-203, *U.S. Statutes at Large* 76 (1953): 400–407.

41. From September 1957 through 1958, the Capelinho volcano erupted causing seismic activity over months and displacing thousands of people in the Azores. The law, passed in September 1958, authorized visas for some of its victims through June 30, 1960. The law was then extended through June 1962. An Act for the Relief of Certain Distressed Aliens, Public Law 85-892, *U.S. Statutes at Large* 72 (1958): 1712–13.

42. The 1965 Immigration and Nationality Act (also known as the Hart-Celler Act) amended the 1952 Immigration and Nationality Act. It allowed for the "conditional entry" of 6 percent of the immigration quota per year (10,200 refugees). It defined refugees as the "oppressed or persecuted, or threatened with oppression or persecution, because of their race, color, religion, national origin, adherence to democratic beliefs, or their opposition to totalitarianism or dictatorship, and . . . persons uprooted by natural calamity or military operations who are unable to return to their usual place of abode." An Act to Amend the Immigration and Nationality Act, Public Law 89-236,

U.S. Statutes at Large 79 (1965), 911–22. For the Senate discussion of refugee status see, for example, U.S. Senate, Subcommittee on Immigration and Naturalization of the Committee of the Judiciary, *Hearings on S. 500 to Amend the Immigration and Nationality Act and for other Purposes, Part I,* 89th Cong., 1st sess., February 10, 24, 25; March 1, 3, 4, 5, 11, 1965 (Washington, D.C.: Government Printing Office, 2010), 131, 173–74, 247, 642.

43. U.S. Senate, Committee on the Judiciary, *Amending the Immigration and Nationality Act and for Other Purposes,* S. Report 748, 89th Cong, 1st sess. (Washington, D.C.: Government Printing Office, 1965), 16.

44. Janet L. Parker, "Victims of Natural Disasters in U.S. Refugee Law and Policy," *Michigan Yearbook of International Legal Studies* 3 (1982): 138.

45. "The island of Faial, Azores experienced earthquakes and volcanic eruptions so intense that a new island came into being, only to be swallowed up by the sea again." Senators John Kennedy of Massachusetts and John Pastore of Rhode Island cosponsored the bill that became law. By 1980, the Census Bureau reported 32,531 foreign-born Azoreans were living in the United States, suggesting that others both predated and followed their migration. "Portuguese Immigrants in the United States," Library of Congress, accessed May 26, 2020, https://www.loc.gov/rr/hispanic/portam/chron6.html. See also "Table 3: World Region and Country or Area of Birth of the Foreign-Born Population: 1960 to 1990," in Campbell Gibson and Kay Jung, "Historical Census Statistics on the Foreign-Born Population of the United States: 1850–2000," Population Division Working Paper No. 81, U.S. Census Bureau, 2006, 39, https://www.census.gov/content/dam/Census/library/working-papers/2006/demo/POP-twps0081.pdf; Michael Doyle, "House Resolution Commemorates the 1958 Azorean Refugees Act," McClatchy DC Bureau website, August 12, 2008, https://www.mcclatchydc.com/news/politics-government/article24495382.html.

46. Parker, "Victims of Natural Disasters in U.S. Refugee Law and Policy," 140.

47. Parker, 141–42. See also Jeanhee Hong, "Refugees of the 21st Century: Environmental Justice," *Cornell Journal of Law and Public Policy* 10, no. 1 (Spring 2001): 323–48, http://scholarship.law.cornell.edu/cgi/viewcontent.cgi?article=1018&context=cjlpp; and Murray and Williamson, "Migration as a Tool for Disaster Recovery," 29.

48. An Act to Amend the Immigration and Nationality Act to Change the Level, and Preference System for Admission of Immigrants to the United States, and to Provide for the Administrative Naturalization, and for Other Purposes, Public Law 101-649, *U.S. Statutes at Large* 104 (1989): 4978–5088; memorandum dated April 29, 2011, American Immigration Council, https://www.americanimmigrationcouncil.org/sites/default/files/general_litigation/Memo_exec_branch_authority.pdf. According to Murray and Williamson, TPS "was a congressional response to the ad hoc use of 'Extended Voluntary Departure' (EVD) as a means of allowing certain populations to remain in the U.S. for limited periods of time." Murray and Williamson, "Migration as a Tool for Disaster Recovery," 36.

49. Ruth Ellen Wasem and Karma Ester, "Temporary Protected Status: Current Immigration Policy and Issues," Congressional Research Service, September 30, 2008, 3, https://digitalcommons.ilr.cornell.edu/cgi/viewcontent.cgi?referer=https://www.google.com/&httpsredir=1&article=1562&context=key_workplace.

50. Madeline Messick and Claire Bergeron, "Temporary Protected Status in the United States: A Grant of Humanitarian Relief That Is Less Than Permanent," *Migration Information Source*, July 2, 2014, http://www.migrationpolicy.org/article/temporary-protected-status-united-states-grant-humanitarian-relief-less-permanent.

51. Before 1990, the attorney general had prosecutorial discretion to allow undocumented nationals of certain countries to remain in the United States under an administrative status called Extended Voluntary Departure. Critics charged that decisions were politically motivated, and in response Congress created TPS. Messick and Bergeron, "Temporary Protected Status in the United States."

52. According to Murray and Williamson, Liberia offers an example. Liberians held TPS from 1991 to 1999. When their TPS expired on September 28, 1999, the Clinton administration granted them DED. Liberia was subsequently redesignated for TPS from 2002 to 2006. When TPS was once again set to expire, the George W. Bush administration granted them DED, which was extended by President Obama. Liberians were once again designated for TPS during 2016–17 because of the Ebola crisis. Muzaffar Chishti, Faye Hipsman, and Sarah Pierce, "Ebola Outbreak Rekindles Debate on Restricting Admissions to the United States on Health Grounds," *Migration Information Source*, October 24, 2014, http://www.migrationpolicy.org/article/ebola-outbreak-rekindles-debate-restricting-admissions-united-states-health-grounds; "Temporary Protected Status Designated Country: Liberia," U.S. Citizenship and Immigration Services (hereafter USCIS), accessed November 30, 2017, https://www.uscis.gov/humanitarian/temporary-protected-status-deferred-enforced-departure/temporary-protected-status-designated-country-liberia; and "Deferred Enforced Departure," USCIS, accessed November 30, 2017, https://www.uscis.gov/humanitarian/temporary-protected-status/deferred-enforced-departure. See also memorandum dated April 29, 2011, American Immigration Council; and Messick and Bergeron, "Temporary Protected Status in the United States."

53. For a discussion of the legal struggle to secure protection for Salvadorans, see Maria Cristina Garcia, *Seeking Refuge: Central American Migration to Mexico, the United States, and Canada* (Berkeley: University of California Press, 2006).

54. Messick and Bergeron, "Temporary Protected Status in the United States."

55. "Termination of the Designation of Montserrat under the Temporary Protected Status Program," USCIS, accessed December 4, 2017, https://www.uscis.gov/ilink/docView/FR/HTML/FR/0-0-0-1/0-0-0-94157/0-0-0-94177/0-0-0-96204.html.

56. Jill H. Wilson, "Temporary Protected Status: Current Immigration and Policy Issues," Congressional Research Service, January 17, 2017, 9, https://fas.org/sgp/crs/homesec/RS20844.pdf.

57. Latin America Digital Beat, "Guatemala Given Little Chance of Temporary Protected Status for Its Citizens in the U.S.," University of New Mexico Digital Repository, May 1, 2008, https://digitalrepository.unm.edu/cgi/viewcontent.cgi?article=10603&context=noticen. See also James Smith, "Guatemala: Economic Migrants Replace Political Refugees," *Migration Information Source*, April 1, 2006, http://www.migrationpolicy.org/article/guatemala-economic-migrants-replace-political-refugees.

58. Messick and Bergeron, "Temporary Protected Status in the United States."

59. U.S. House of Representatives, Subcommittee on Immigration and Claims of the Committee on the Judiciary, *Designations of Temporary Protected Status and Fraud in Prior Amnesty Programs*, 106th Cong, 1st sess., March 4, 1999 (Washington, D.C.: Government Printing Office, 2000), 9.

60. Norwegian Refugee Council, *Climate Changed*; Bill Frelick and Barbara Kohnen, "Filling the Gap: Temporary Protected Status," *Journal of Refugee Studies* 8, no. 4 (1995): 339–63; U.S. House of Representatives, Subcommittee on Immigration and Claims of the Committee on the Judiciary, *Designations of Temporary Protected Status and Fraud in Prior Amnesty Programs*, 101–4. To alleviate some of these concerns, the Lawyers' Committee for Human Rights urged the attorney general to publish guidelines explaining the circumstances under which TPS designation might be appropriate, as well as an assessment of the likely duration of the displacement, the ability of the United States to play a role in resolving the underlying conflict, and the scope of the humanitarian need.

61. Several countries offer discretionary "humanitarian stays" for people displaced by environmental disasters. Finland and Sweden, for example, offer either temporary or permanent protection to foreign nationals who cannot return safely to their home country because of an "environmental disaster." Norwegian Refugee Council, *Climate Changed*; Frelick and Kohnen, "Filling the Gap."

62. Legal scholar David A. Martin, for example, has recommended that the State Department broaden the class of persons eligible to be admitted through the U.S. Refugee Admissions Program without "individually applying the Convention refugee definition." Similarly, the 2010 Refugee Protection Act, introduced by Senator Patrick Leahy of Vermont, proposed broadening access to the Refugee Admissions Program to those who have "a need for resettlement due to vulnerabilities or a lack of local integration prospects in their country of first asylum." At the time of publication, neither Martin's recommendations nor Leahy's various refugee reform bills have been adopted. David A. Martin, "The United States Refugee Admissions Program: Reforms for a New Era of Refugee Resettlement," Migration Policy Institute, May 2005, http://www.migrationpolicy.org/research/united-states-refugee-admissions-program-reforms-new-era-refugee-resettlement; Refugee Protection Act of 2010, S. 3113, 111th Cong., https://www.govtrack.us/congress/bills/111/s3113. Previous and subsequent revisions of this bill did not include this language. See also "Top 10 of 2009."

63. Murray and Williamson, "Migration as a Tool for Disaster Recovery," 33.

64. Murray and Williamson, 1. The UNHCR also noted that "refugees and asylum seekers, especially those in the industrialized states, remit significant amounts of money to household and community members, both in countries of origin and to refugees in other asylum countries." See "Refugee Protection and International Migration," UNHCR, accessed February 21, 2018, 6–7, http://www.unhcr.org/protection/migration/4a24efoca2/unhcr-refugee-protection-international-migration.html; Charles Kenny, "The Haitian Migration," *Foreign Policy*, January 9, 2012, http://foreignpolicy.com/2012/01/09/the-haitian-migration/; and Kathleen Newland, Aaron Terrazas, and Roberto Munster, *Diaspora Philanthropy: Private Giving and Public Policy* (Washington, D.C.: Migration Policy Institute, September 2010), https://www.migrationpolicy.org/pubs/diasporas-philanthropy.pdf.

65. "Remittance Flows Worldwide in 2015," Pew Research Center, August 31, 2016, http://www.pewglobal.org/interactives/remittance-map/. The challenge for policy makers, as Laczko and Collett have noted, is to "convert this desire and ability to contribute to home country reconstruction and development [on the part of diasporic communities] into longer-term, sustainable projects." See Laczko and Collett, "Assessing the Tsunami's Effects on Migration."

66. Department of Homeland Security, *Written Testimony of FEMA Region 9 Administrator Robert Fenton for a House Committee on Transportation and Infrastructure, Subcommittee on Economic Development, Public Buildings, and Emergency Management Hearing Titled "Impacts of the 2017 Wildfires in the United States,"* March 20, 2018, https://www.dhs.gov/news/2018/03/20/written-testimony-fema-house-transportation-and-infrastructure-subcommittee-economic. This assessment appears in several FEMA reports and testimonies. See, for example, FEMA, "Statement of William B. Long before the Committee on Oversight and Government Reform, House of Representatives," November 29, 2018, https://www.fema.gov/sites/default/files/2020-07/william-b-long_disaster-response-recovery_statement_11-29-2018.pdf.

Chapter 1

1. According to the British Library's Endangered Archives Programme, volcanic activity and "inappropriate storage and handling" resulted in the destruction of most of the country's archival storage facilities, "creating an urgent need for proper documentation and storage." The British Archives is currently identifying archival material held throughout Montserrat to "assess its condition and prepare a long-term plan for its safe storage, digitisation and increased public access and awareness of this endangered resource." See "Montserrat in Written Records and Photographs: Preserving the Archive for the Nation and the Montserrat Diaspora," British Library Endangered Archives Programme, accessed March 14, 2021, https://eap.bl.uk/project/EAP769.

2. Estimates of the Montserratian population vary. The UK Department for International Development placed the population at over 10,000. See Edward Clay et al., *An Evaluation of HMG's Response to the Montserrat Volcanic Emergency*, Evaluation Report EV635, 1, no. 1, (December 1999), https://www.gov.uk/government/uploads/system/uploads/attachment_data/file/67966/ev635.pdf. See also B. P. Kokelaar, "Setting, Chronology, and Consequences of the Eruption of Soufrière Hills Volcano, Montserrat (1995–1999)," in *The Eruption of Soufrière Hills Volcano, Montserrat, from 1995 to 1999*, ed. Timothy H. Druitt and B. P. Kokelaar (London: Geological Society, 2002), 1–43.

3. "Extension of the Designation of Montserrat under the Temporary Protected Status Program," *Federal Register* 66 (August 3, 2001): 40834–5, https://www.govinfo.gov/content/pkg/FR-2001-08-03/pdf/01-19475.pdf.

4. Howard Fergus, *Montserrat: History of a Caribbean Colony* (London: Macmillan, 1994); Howard Fergus, "Montserrat's Days of Lime and Cotton," *Caribbean Quarterly* 28, no. 2 (1982): 10–18.

5. Jerome L. McElroy and Klaus de Albuquerque, "The Economic Impact of Retirement Tourism in Montserrat: Some Provisional Evidence," *Social and Economic Studies* 41 (June 1992): 127–52.

6. By 1648, there were 1,000 Irish families on the island, from Ireland, Barbados, and Virginia. Montserratians have retained some Irish customs and cuisine. St. Patrick's Day is a national holiday in Montserrat, although some residents have advocated that the island commemorate instead the date of a failed slave insurrection that occurred on March 17. Diane McDougall-Tierney, "Montserrat: The Way the Caribbean Used to Be," *Toronto Star*, September 26, 1992, Nexis Uni; Michael Jarvis, "Montserrat's St. Patrick's Day: Irish Connection or African Disconnection?," *Caribbean News Now*, March 18, 2019, Nexis Uni.

7. The AIR Studio shut its doors in 1989, after Hurricane Hugo, and is the subject of a 2021 documentary film called *Under the Volcano* (directed by Gracie Otto); see IMDB entry at https://www.imdb.com/title/tt9598492/. Ryan Schuessler, "George Martin's AIR Studio Is a Symbol of Montserrat's Lost Era as Pop Music Hub," *The Guardian*, January 28, 2016.

8. Secretary of State for Foreign and Commonwealth Affairs, *Partnership for Progress and Prosperity, Britain and the Overseas Territories*, 1999, 60, http://www.ukotcf.org/pdf/charters/WhitePaper99full.pdf. In 2011, the office of the chief minister was replaced by the office of the premier. The governor no longer has sole discretionary powers.

9. The fourteen British Overseas Territories (BOTs) are Anguilla, Bermuda, the British Virgin Islands, the Cayman Islands, the Falkland Islands, Gibraltar, Montserrat, the Pitcairn Islands, St. Helena and its dependencies Ascension and Tristan da Cunha, the Turks and Caicos Islands, the British Antarctic Territory, the British Indian Ocean Territory, and South Georgia and the South Sandwich Islands. The "Sovereign Base Areas" of Akrotiri and Dhekelia in Cyprus are considered two additional overseas territories. In 2010, the combined population of the BOTs was approximately 200,000 people, with Bermuda having the largest population. The BOTs are supported by the United Kingdom Overseas Territories Association, which acts as a lobbying group in London to promote the common interests of the BOTs. Secretary of State for Foreign and Commonwealth Affairs, *Partnership for Progress and Prosperity*, 8.

10. Frank A. Perret, *The Volcano-Seismic Crises at Montserrat, 1933–1937* (Washington, D.C.: Carnegie Institution of Washington, 1939), 47; Nigel Sadler, "Montserrat in Written Records and Photographs: Preserving the Archive for the Nation and the Montserrat Diaspora," British Library Endangered Archives Programme, 7, 11, accessed June 1, 2021, https://eap.bl.uk/sites/default/files/EAP769%20survey.pdf. The pilot project to preserve Montserrat's archival heritage was essential not just for Montserratians on the island but for the Montserratian diaspora. "Montserrat now has a vast Diaspora, with over 2/3rds of people born in Montserrat now living overseas, mostly in the U.K. but also spread throughout the Caribbean," wrote Sadler. "This has disconnected many thousands of people from their place of birth and from their records. At the same time the people who have left Montserrat have done so often taking all their possessions, including important family archives, removing them from a country that has recently been more defined by a volcano rather than by its people."

11. "Deadliest Eruption," Volcano World, Oregon State University, accessed January 4, 2018, http://volcano.oregonstate.edu/deadliest-eruption; "Soufrière St. Vincent," Volcano World, Oregon State University, accessed January 4, 2018, http://volcano

.oregonstate.edu/soufri%C3%A8re-st-vincent. See also Timothy H. Druitt and B. P. Kokelaar, eds., *The Eruption of Soufrière Hills Volcano, Montserrat, from 1995 to 1999* (London: Geological Society, 2002), x. In 2021, a soufrière on the island of St. Vincent erupted requiring the evacuation of 16,000 people.

12. "Guadeloupe Volcano Expected to Erupt; 72,000 Evacuated," *New York Times*, August 16, 1976, http://www.nytimes.com/1976/08/16/archives/guadeloupe-volcano-expected-to-erupt-72000-evacuated-guadeloupe.html.

13. McDougall-Tierney, "Montserrat."

14. W. P. Aspinall et al., "The Montserrat Volcano Observatory: Its Evolution, Organization, Role and Activities," in Druitt and Kokelaar, *Eruption of Soufrière Hills Volcano*, 211; "Volcano Keeps Montserrat in Constant State of Alert; Caribbean Isle Is Full of Dramatic Reminders of Destruction," *Washington Post*, January 21, 2001, https://www.washingtonpost.com/archive/politics/2001/01/21/volcano-keeps-montserrat-in-constant-state-of-alert/3a400be0-027a-449b-8063-7cf180048b45/; "Caribbean Volcano Provides Scientists Valuable Lessons," *Washington Post*, March 11, 2001, https://www.washingtonpost.com/archive/politics/2001/03/11/caribbean-volcano-provides-scientists-valuable-lessons/4a8226bd-01ac-4af3-b296-dffd16d385df/. The Montserrat volcano was one of ninety-two active volcanoes studied by scientists.

15. The Montserrat risk assessment was prepared in 1987 as part of the Pan-Caribbean Disaster Prevention and Preparedness Project but was overlooked in the immediate aftermath of the emergency. Clay et al., *Evaluation of HMG's Response to the Montserrat Volcanic Emergency*, 2, 9; Sian Clare, "Short Tells of Kafka-Like Prison Plan for Volcano Island," Press Association, May 5, 1998, Nexis Uni.

16. Kokelaar, "Setting, Chronology, and Consequences," 5; Myrnon W. Chin and W. H. E. Suite, *Hurricane Hugo: A Survey of Damage in Montserrat and Antigua* (Trinidad: National Emergency Management Agency, January 11, 1989), 3, http://www.oecs.org/ogu-resources/hurricane-hugo-a-survey-of-d1999amage-in-montserrat-and-antigua-november-1989-55660pt-pdf (page discontinued); "Montserrat Hurricane—September 1989," International Rescue Corps, accessed January 8, 2018, http://www.intrescue.info/hub/index.php/missions/montserrat-hurricane-sept-1989/; "Montserrat Aid Programme Comes under Fire from U.K. Watchdog," *The Guardian*, July 16, 2013, Nexis Uni; Tracey Skelton, "Political Uncertainties and Natural Disasters: Montserratian Identity and Colonial Status," *Interventions* 2 (2000): 103–17.

17. Polly Pattullo, *Fire from the Mountain: The Tragedy of Montserrat and the Betrayal of Its People* (London: Constable Press, 2000), 61.

18. Kokelaar, "Setting, Chronology, and Consequences," 5, 20.

19. Clay et al., *Evaluation of HMG's Response to the Montserrat Volcanic Emergency*, 2.

20. Kokelaar, "Setting, Chronology, and Consequences," 1.

21. Clay et al., *Evaluation of HMG's Response to the Montserrat Volcanic Emergency*, 3–4.

22. These included representatives from the British Geological Survey, the Seismic Research Unit of the University of the West Indies in Trinidad, and the U.S. Geological Survey Volcanic Crisis Assistance Team. Today, the Montserrat Volcano Observatory is operated by the Seismic Research Unit and the British Geological Survey, with funding from the UK Overseas Development Administration and the Government

of Montserrat. See "Montserrat Volcano Observatory," Michigan Tech Volcano page, Geological and Mining Engineering and Sciences, Michigan Technological University, accessed April 4, 2021, http://www.geo.mtu.edu/volcanoes/west.indies/soufriere/govt/miscdocs/mvo.html.

23. S. C. Loughlin et al., "Eyewitness Accounts of the 25 June 1997 Pyroclastic Flows and Surges at Soufrière Hills Volcano, Montserrat, and Implications for Disaster Mitigation," in Druitt and Kokelaar, *Eruption of Soufrière Hills Volcano*, 211.

24. Loughlin et al., 211. According to journalist Polly Pattullo, it was easier to convince farmers on the western flanks to evacuate because the winds deposited ash everywhere, burning fields and suffocating people and livestock. See Pattullo, *Fire from the Mountain*, 8. Geographer Tracey Skelton explained: "[The] people from the eastern villages of the island had long earned or supplemented their incomes from farming and livestock rearing. Their daily lives were spent out on their land, tending crops and animals. Many of these people adapted poorly to the claustrophobic and cramped conditions in the shelters and so walked across many miles of thick brush to return to their land in the east, many residing in what remained of their own homes." Tracey Skelton, "Globalizing Forces and Natural Disaster: What Can Be the Future for the Small Caribbean Island of Montserrat?," in *Globalization: Theory and Practice*, 2nd ed., ed. Eleonore Kofman and Gillian Youngs (London: Continuum, 2003), 69.

25. Skelton, "Globalizing Forces and Natural Disaster," 70.

26. Pattullo, *Fire from the Mountain*, 6–7.

27. Skelton, "Globalizing Forces and Natural Disaster," 69; Action by Churches Together International, "Montserrat Volcanic Eruption Appeal LAMN71," ReliefWeb, July 17, 1997, https://reliefweb.int/report/montserrat/montserrat-volcanic-eruption-appeal-lamn71.

28. G. C. Mayberry et al., "Dynamics of Volcanic and Meteorological Clouds Produced on 26 December (Boxing Day) 1997 at Soufrière Hills Volcano, Montserrat," in Druitt and Kokelaar, *Eruption of Soufrière Hills Volcano*, 540; Clay et al., *Evaluation of HMG's Response to the Montserrat Volcanic Emergency*, 74, 76.

29. Most territories are said to be self-sufficient, but the DFID maintains development assistance programs in St. Helena, Montserrat, and Pitcairn. "United Kingdom: Investment Climate Statement 2010," *The Telegraph*, January 15, 2010, https://www.telegraph.co.uk/news/wikileaks-files/london-wikileaks/8304871/United-Kingdom-Investment-Climate-Statement-2010.html.

30. Pattullo, *Fire from the Mountain*, 94.

31. Skelton, "Globalizing Forces and Natural Disaster," 67.

32. "Climatology of Caribbean Hurricanes—Montserrat," Caribbean Hurricane Network, accessed January 14, 2021, https://stormcarib.com/climatology/TKMN_dec_atl.htm; Ian Black, "Volcano 'Fiasco Britain's Fault,'" *The Guardian*, October 29, 1997, Nexis Uni.

33. Skelton, "Political Uncertainties and Natural Disasters," 112. According to Skelton, Montserratians had few ways to publicly express their sentiments and opinions except through this radio station: "Telephone debates [were] lively, angry, and humourous," wrote Skelton. Skelton, "Globalizing Forces and Natural Disaster," 69.

NOTES TO PAGES 27–29

34. Kim Segupta, "Controversies of Claire," *The Independent*, October 9, 1998, http://www.independent.co.uk/news/controversies-of-clare-short-1177162.html; Pattullo, *Fire from the Mountain*, 108–10.

35. "World News in Brief: Diplomatic Row in Montserrat," *The Guardian*, October 29, 1997, Nexis Uni.

36. Skelton, "Political Uncertainties and Natural Disasters," 111; Skelton, "Globalizing Forces and Natural Disaster," 73.

37. David S. Brandt, "Letter: Relations with Montserrat," *The Guardian*, October 31, 1997, 20, Nexis Uni.

38. Skelton, "Globalizing Forces and Natural Disaster," 68.

39. Black, "Volcano 'Fiasco Britain's Fault'"; Ian Black, "Caring Britain Pledges to Help Rebuild Montserrat," *The Guardian*, September 16, 1997, Nexis Uni; Polly Pattullo, "Under the Volcano: The British Government's Inadequate Response to the Crisis in Montserrat Is Criticised in a New Report," *The Guardian*, May 5, 2000, Nexis Uni; Ian Black, "Montserrat Chief Attacks Britain for Spurring Exodus," *The Guardian*, January 17, 1998, Nexis Uni.

40. According to *The Guardian*, "The Montserrat disaster highlighted a turf war between the Foreign Office and DFID. Before the general election, DFID's predecessor, the Overseas Development Agency, had been part of the Foreign Office, and there was resentment in the Foreign Office when it was made a separate department." Ewen Macaskill, "Ministers Caught in Montserrat Fallout; A Report by MPs Blames Cook and Short for the Shambolic Handling of the Island Volcano Disaster," *The Guardian*, August 5, 1998, 3, Nexis Uni.

41. Clay et al., *Evaluation of HMG's Response to the Montserrat Volcanic Emergency*, 1.

42. Clay et al., 1–4.

43. Clay et al., 1.

44. Fifty-two witnesses testified at the juried inquest. Bennette Roach, "As We Remember June 25, 1997," *Montserrat Reporter*, June 19, 2015, https://www.themontserratreporter.com/as-we-remember-june-25-1997/; "UK Blamed for Volcano Deaths," *BBC News*, January 12, 1999, http://news.bbc.co.uk/2/hi/uk_news/253237.stm; Pattullo, *Fire from the Mountain*, 182–83.

45. Clay et al., *Evaluation of HMG's Response to the Montserrat Volcanic Emergency*, 21; Secretary of State for Foreign and Commonwealth Affairs, *Partnership for Progress and Prosperity*, 33; "British Ship Prepares to Evacuate Island," *Washington Post*, August 20, 1997, https://www.washingtonpost.com/archive/politics/1997/08/20/british-ship-prepares-to-evacuate-island/66ca00c9-7052-4044-be75-9dc976e0e0f7/.

46. According to Tracey Skelton, the government of Antigua "received the waiving of aid/loan repayments due to the British Government of £1.25 million; a development grant of £3 million for health and education projects which replaced a scheduled aid loan of £2 million; and £1 million in grants for the development of small- and medium-sized enterprises for the Montserratian Community in Antigua." Skelton, "Political Uncertainties and Natural Disasters," 111. See also Pattullo, *Fire from the Mountain*, 180.

47. Skelton, "Political Uncertainties and Natural Disasters," 105.

48. Secretary of State for Foreign and Commonwealth Affairs, *Partnership for Progress and Prosperity.*

49. "Types of British Nationality," Gov.UK, accessed January 6, 2018, https://www.gov.uk/types-of-british-nationality/british-overseas-territories-citizen.

50. According to Skelton, there was little support for independence on the island. Skelton, "Political Uncertainties and Natural Disasters," 103, 116.

51. Tim MacDonald, "Volcano Keeps Montserrat in Constant State of Alert; Caribbean Isle Is Full of Dramatic Reminders of Destruction," *Washington Post*, January 21, 2001, A19, https://www.washingtonpost.com/archive/politics/2001/01/21/volcano-keeps-montserrat-in-constant-state-of-alert/3a400be0-027a-449b-8063-7cf180048b45/.

52. "Extension of the Designation of Montserrat under the Temporary Protected Status Program," *Federal Register* 67 (August 3, 2002): 47022–24, https://www.federalregister.gov/documents/2002/07/17/02-18040/extension-of-the-designation-of-montserrat-under-the-temporary-protected-status-program.

53. "Termination of the Designation of Montserrat under the Temporary Protected Status Program," *Federal Register* 69 (July 6, 2004): 40642–5, https://www.federalregister.gov/documents/2004/07/06/04-15243/termination-of-the-designation-of-montserrat-under-the-temporary-protected-status-program-extension.

54. "Caricom Countries Ready to Assist Volcano Ravaged Montserrat," *BBC Monitoring—Latin America*, July 16, 2003, Nexis Uni.

55. In 1995 Montserrat's population registered at 9,848, but by 2005 the population registered at 4,835. The largest population size registered in the post–World War II era was 14,233 in 1955. See "Montserrat Population," World Population Review, accessed June 21, 2020, https://worldpopulationreview.com/countries/montserrat-population/.

56. "Extension of the Designation of Montserrat under the Temporary Protected Status Program," July 1, 2003.

57. Mark Fineman, "Devastated Islands Rebuild: Internal Political Struggles, Rows with British Bureaucrats Slow Montserrat's Recovery," *Montreal Gazette*, November 13, 1999, B3, Nexis Uni.

58. "Hazard Level," Montserrat Volcano Observatory, accessed January 18, 2018, http://www.mvo.ms/.

59. According to Stuart B. Philpott, migration at the lower end of the class scale was more likely to be temporary and a means to improve one's station in life. Upper-class expatriates were more likely to settle permanently outside of the island. Stuart B. Philpott, *West Indian Migration: The Montserrat Case* (London: Athlone Press, 1973), 69, 167, 170, 181, 196–97.

60. The McCarran-Walter Act of 1952 denied Afro-Caribbean peoples the right to take advantage of the quotas set for Great Britain. Violet Showers Johnson discusses Montserratians in Boston in her book *The Other Black Bostonians: West Indians in Boston, 1900–1950* (Bloomington: Indiana University Press, 2006).

61. "Montserrat Population."

62. Gertrude Shotte, "Diasporic Transnationalism: Relocated Montserratians in the U.K.," *Caribbean Quarterly* 53 (2007): 42.

63. Pattullo, *Fire from the Mountain*, 102; Phil Davison, "Island Declares 'War' on Britain," *The Independent*, August 25, 1997, Nexis Uni.

64. Clay et al., *Evaluation of HMG's Response to the Montserrat Volcanic Emergency*, 1; Secretary of State for Foreign and Commonwealth Affairs, *Partnership for Progress and Prosperity*, 61; Frank Mills, "The Transnationalization of Immigration Policies and Its Effects on Caribbean Microstates," *Wadabagei: A Journal of the Caribbean and Its Diaspora* 6 (2003): 1–42.

65. "Nowhere to Call Home: Montserratians Still Struggling to Settle after Volcano Devastation," *Nottingham Evening Post*, June 7, 1999, Nexis Uni; "A Long Way from Home, Traumatic Times," *Nottingham Evening Post*, June 4, 1999, Nexis Uni.

66. "Nowhere to Call Home."

67. "Canada's Immigration Policy Generous with Montserratians," *Toronto Star*, September 9, 1997, A18, Nexis Uni.

68. Montserrat was the only British Overseas Territory that is a full member of the Caribbean community. "CARICOM Governments Pledge More Assistance to Montserrat," CARICOM, July 14, 1997, https://caricom.org/caricom-governments-pledge-more-assistance-to-montserrat/; Adventist Development and Relief Agency International, "ADRA Continues Relief in Montserrat," ReliefWeb, October 14, 1997, https://reliefweb.int/report/montserrat/adra-continues-relief-montserrat.

69. US Department of Justice, "Table 3: Immigrants Admitted by Region and Country of Birth Fiscal Years 1989–99," *1999 Statistical Yearbook of the Immigration and Naturalization Service*, March 2002, 27, https://www.dhs.gov/sites/default/files/publications/Yearbook_Immigration_Statistics_1999.pdf; New York City Department of Urban Planning, Population Division, "The Newest New Yorkers 2000," NYC Department of City Planning website, October 2004, 200–201, 208–9, 218–19, 226 27, https://www1.nyc.gov/site/planning/planning-level/nyc-population/newest-new-yorkers-2000.page. The national figures reported in the New York City report (1,092) differ from those in the *Statistical Yearbook*. According to the NYC Department of Urban Planning, 93 of those who settled in New York City between 1990 and 1999 arrived through the "employment preference," while 377 arrived as part of family reunification (or family preference) visas. According to this report, 1,103 Montserratians immigrated to the United States between 1982 and 1989, 976 through the family preference category, and 405 Montserratians settled in NYC between 1982 and 1989.

70. "Countries of Birth for U.S. Immigrants, 1990," Migration Policy Institute, accessed January 14, 2021, https://www.migrationpolicy.org/programs/data-hub/charts/immigrants-countries-birth-over-time; Skelton, "Globalizing Forces and Natural Disaster," 68; Larry Fish, "As Antigua Gets Volcano Evacuees, Britain Is Faulted," *Philadelphia Inquirer*, September 1, 1997, Nexis Uni; Tara George, "Immigs Reach Out to Montserratians," *New York Daily News*, September 3, 1997, 3, Nexis Uni.

71. Individuals with no nationality but whose country of last habitual residence was Montserrat also qualified for TPS. "Designation of Montserrat under Temporary Protected Status," *Federal Register* 62 (August 28, 1997): 45685. Most journalists writing about Montserratians stated that 292 Montserratians received TPS, but USCIS figures list 325. See, for example, Maia Jachimowicz, "Visitor Tracking Technology Faces

Obstacles," *Migration Information Source*, October 1, 2003, https://www.migrationpolicy.org/article/visitor-tracking-technology-faces-obstacles.

72. "Extension of Designation of Montserrat under Temporary Status Protection Program," *Federal Register* 65 (August 28, 2000): 58806–8, https://www.federalregister.gov/documents/2000/10/02/00-25250/extension-of-designation-of-montserrat-under-temporary-protected-status-program.

73. "Termination of the Designation of Montserrat under the Temporary Protected Status Program; Extension of Employment Authorization Documentation," *Federal Register* 69 (July 6, 2004): 40642–45, https://www.federalregister.gov/documents/2004/07/06/04-15243/termination-of-the-designation-of-montserrat-under-the-temporary-protected-status-program-extension.

74. "Britain Discusses New Policy Regarding Montserratians Residing in the USA," *BBC Summary of World Broadcasts*, August 4, 2004, Nexis Uni.

75. Joann Fitzpatrick, "Our Opinion: Shameful Bureaucratic Catch-22," *Patriot Ledger* (Quincy, Mass.), February 9, 2005, 22, Nexis Uni.

76. "Volcanic Absurdity," *Washington Post*, August 15, 2004, B06, https://www.washingtonpost.com/archive/opinions/2004/08/15/volcanic-absurdity/d4f7e9c0-2dad-4fc0-9b08-b9fa30e30517/.

77. Ruth Ellen Wasem reported that 1,000 Montserratians held TPS in 1999 based on data from the Immigration and Naturalization Service for designated status or work authorizations. This was an approximate number and did not include others who might be eligible but did not avail themselves of this status. Ruth Ellen Wasem, "Immigration: Temporary Protected Status Background and Issues, Report No. 98-759," Congressional Research Service, January 5, 1999, 4.

78. Jachimowicz, "Visitor Tracking Technology Faces Obstacles"; Nina Bernstein, "U.S. Is Ending Haven for Those Fleeing a Volcano," *New York Times*, August 9, 2004, http://www.nytimes.com/2004/08/09/nyregion/us-is-ending-haven-for-those-fleeing-a-volcano.html. For the number of Montserratians on TPS, see Ruth Ellen Wasem and Karma Ester, "Temporary Protected Status: Current Immigration Policy and Issues," Congressional Research Service, November 4, 2004, https://www.everycrsreport.com/files/20041104_RS20844_d570c76458155fd585adc5eb2a486faff18b5f2d.pdf.

79. Bennette Roach, "U.S. Government Ends Special Protected Status for Montserratians," Associated Press, February 27, 2005, Nexis Uni.

80. "Montserratians in USA Continue to Appeal for Temporary Protected Status," *BBC Monitoring Latin America*, February 15, 2005, Nexis Uni.

81. To Provide for Adjustment of Immigration Status for Certain Aliens Granted Temporary Protected Status in the United States Because of Conditions in Montserrat, H.R. 1726, 107th Cong. (2001), https://www.govtrack.us/congress/bills/107/hr1726; To Provide for Adjustment of Immigration Status for Certain Aliens Granted Temporary Protected Status in the United States Because of Conditions in Montserrat, H.R. 603, 108th Cong. (2003), https://www.govtrack.us/congress/bills/108/hr603; To Provide for Adjustment of Immigration Status for Certain Aliens Granted Temporary Protected Status in the United States Because of Conditions in Montserrat, H.R. 342, 109th Cong. (2005), https://www.govtrack.us/congress/bills/109/hr342.

NOTES TO PAGES 36–39

82. "Congressman Lynch Calls on President Bush to Allow Montserratian Refugees to Stay in America," press release, February 8, 2005, https://lynch.house.gov/press-release/congressman-lynch-calls-president-bush-allow-montserratian-refugees-stay-america.

83. Montserrat Immigration Fairness Act, S. 2816, 108th Cong. (2004), https://www.govtrack.us/congress/bills/108/s2816; Montserrat Immigration Fairness Act, S. 297, 109th Cong. (2005), https://www.govtrack.us/congress/bills/109/s297.

84. "Schumer Urges DHS to Grant Montserratians Deferred Enforced Departure Allows Them Another 12–18 Months to Get Affairs In Order," press release, August 2, 2006, https://www.schumer.senate.gov/newsroom/press-releases/schumer-urges-dhs-to-grant-montserratians-deferred-enforced-departure-allows-them-another-12-18-months-to-get-affairs-in-order.

85. "Bush Administration Denies Temporary Protected Status Extension for Montserratians," *U.S. Fed News*, February 25, 2005, Nexis Uni.

86. Bennette Roach, "Montserratians in England Are Still Getting Little, and Confusing Information," *Montserrat Reporter*, May 7, 1999, http://www.montserratreporter.org/news0599-1.htm; "Government of Montserrat's Briefing Note for Returning Montserratians," *Montserrat Reporter*, accessed January 30, 2018, http://www.montserratreporter.org/return_brief.htm.

87. The other two maritime exclusion zones are located on the western side of the island. The most southerly of the two extends for two kilometers offshore and the third for a half kilometer offshore. For more information, see "Montserrat," Government of the United Kingdom, accessed June 15, 2020, https://www.gov.uk/foreign-travel-advice/montserrat/natural-disasters.

88. Clay et al., *Evaluation of HMG's Response to the Montserrat Volcanic Emergency*, 75.

89. Skelton, "Globalizing Forces and Natural Disaster," 71, 100. According to Skelton, an estimated 12 percent of Montserratians who had property in the exclusion zone were still paying mortgages for homes they would never again inhabit.

90. "Situation Analysis of Children in Montserrat" (Barbados, 2016), 26, UNICEF, Office for the Eastern Caribbean Area, https://www.unicef.org/easterncaribbean/Montserrat_SitAN_2016_WEB.pdf. For a discussion of standards of living, see Secretary of State for Foreign and Commonwealth Affairs, *Partnership for Progress and Prosperity*, 30.

91. Fineman, "Devastated Islands Rebuild."

92. IOM, "How to Help Migrants Integrate Using Disaster Recovery and Preparedness," IOM Regional Office for Central America, North America and the Caribbean website, accessed January 16, 2021, https://rosanjose.iom.int/SITE/en/blog/how-help-migrants-integrate-using-disaster-recovery-and-preparedness?page=15.

93. Caribbean Disaster Emergency Response Agency, "Situation Update on the Montserrat Volcanic Emergency," ReliefWeb, September 22, 1998, https://reliefweb.int/report/montserrat/situation-update-montserrat-volcanic-emergency; IOM, "How to Help Migrants Integrate Using Disaster Recovery and Preparedness."

94. Patrik Henriksson, "The Value of Citizenship in a British Overseas Territory: Formal and Substantive British Citizenship in Montserrat" (thesis, Linnaeus University, Växjö, Sweden, 2018), 17–18.

95. Polly Pattullo, "After the Volcano: Ten Years Ago Today Came the First Warnings of the Volcanic Blast That Was to Devastate the Tiny Caribbean Island of Montserrat," *The Guardian*, July 18, 2005, Nexis Uni.

96. Pattullo, "After the Volcano."

97. Fineman, "Devastated Islands Rebuild."

98. Montserrat does not appear on the World Bank's charts of remittance data, for example. See "Migration and Remittances Data," World Bank, November 16, 2017, http://www.worldbank.org/en/topic/migrationremittancesdiasporaissues/brief/migration-remittances-data; and IOM, "How to Help Migrants Integrate Using Disaster Recovery and Preparedness." Polly Pattullo lists the Montserrat People's Progressive Alliance as one organization that raised funds. See Pattullo, *Fire from the Mountain*, 132.

99. UK Foreign and Commonwealth Office, *The Overseas Territories: Security, Success, and Sustainability*, June 2012, 17, https://www.gov.uk/government/uploads/system/uploads/attachment_data/file/14929/ot-wp-0612.pdf; Secretary of State for Foreign and Commonwealth Affairs, *Partnership for Progress and Prosperity*, 34.

100. Skelton, "Globalizing Forces and Natural Disaster," 75.

101. Government of Montserrat, *Montserrat Sustainable Development Plan 2008–2020*, accessed June 15, 2020, https://www.gov.ms/wp-content/uploads/2020/08/Montserrat-SDP-2008-to-2020.pdf.

102. Government of Montserrat, *Physical Development Plan for North Montserrat, 2012–2022*, April 2012, http://www.gov.ms/wp-content/uploads/2018/11/Physical-Development-Plan.pdf. See also "Montserrat Government to Outline 10-Year Development Plan," Caribbean Media Corporation News Agency, October 18, 2011, Nexis Uni; "Grant Funding of £14.4 Million to Develop Port at Little Bay, Montserrat," Caribbean Development Bank, December 29, 2017, http://wp.caribbeannewsnow.com/2017/12/29/grant-funding-14-4-million-develop-port-little-bay-montserrat/.

103. Office of the Premier, Diaspora Affairs, *Guide for Montserrat Nationals Returning Home*, 2010, https://www.developingmarkets.com/sites/default/files/Diaspora%20re-entry%20guide.pdf.

104. One watchdog group, the Independent Commission for Aid Impact, wrote, "With Antigua only 43 kilometers away, ferry access could have been a more suitable alternative. Many islands with small populations cope well with this means of access and accept that adverse weather conditions may mean that access (whether by ferry or air) is not always possible." Montserrat needed to have "realistic expectations and achieve self-sufficiency." "Montserrat Aid Programme Comes under Fire from UK Watchdog," *The Guardian*, July 16, 2013, Nexis Uni.

105. "Montserrat: The 'Caribbean Pompeii' Where the Rolling Stones Recorded, Hopes to Rise from Its Volcanic Ashes," *South China Morning Post*, October 7, 2019, https://www.scmp.com/magazines/post-magazine/travel/article/3031191/montserrat-caribbean-pompeii-where-rolling-stones. For other examples see Montserrat Tourism Division web page, accessed June 24, 2020, http://www.visitmontserrat.com/places-to-see/; Elissa Gray, "Why Montserrat Is This Year's Hottest Caribbean Cruise Destination," *Conde Nast Traveler*, July 7, 2016, https://www.cntraveler.com/stories/2016-07-07/why-montserrat-is-this-years-hottest-caribbean-cruise-destination; "Visiting Montserrat's Soufrière Hills," Caribbean & Co. website, May

27, 2015, https://www.caribbeanandco.com/visiting-montserrats-soufriere-hills-volcano/; Montserrat Island Tours web page, accessed January 16, 2018, http://montserratislandtours.com/; Laura Chubb, "Dark Tourism: Why It's Ok to Visit Disaster Zones on Holiday," *The Independent*, October 11, 2016, http://www.independent.co.uk/travel/news-and-advice/disaster-tourism-new-orleans-auschwitz-hiroshima-hurricane-katrina-fukushima-killing-fields-a7356156.html; and Mark Rogers, "Mysterious Montserrat: Volcano-Buried City, Beatles Legacy," *USA Today*, July 5, 2016, https://www.usatoday.com/story/travel/experience/caribbean/2016/06/28/montserrat/86437812/.

106. "Montserrat: The 'Caribbean Pompeii'"; "Montserrat Records Highest Tourist Arrival Figures since Eruption of Volcano," *CANA News*, January 29, 2020, Nexis Uni; "Number of International Tourist Arrivals in Montserrat between 2005 and 2018," Statista, accessed June 21, 2020, https://www.statista.com/statistics/816472/montserrat-number-of-tourist-arrivals/.

107. Ryan Schuessler, "'Ash to Cash': Montserrat Gambles Future on the Volcano That Nearly Destroyed It," *The Guardian*, January 28, 2016, https://www.theguardian.com/world/2016/jan/28/montserrat-volcano-british-territory-geothermal-energy-tourism-sand-mining; Graham Alexander Ryan, "How One Island Powered Itself with a Volcano," *Washington Post Blog*, July 14, 2014, https://www.washingtonpost.com/posteverything/wp/2014/07/13/how-one-island-powered-itself-with-a-volcano/.

108. "Montserrat Population."

109. Graeme T. Swindles, Elizabeth J. Watson, Ivan P. Savov, Ian T. Lawson, Anja Schmidt, Andrew Hooper, Claire L. Cooper, et al., "Climatic Control on Icelandic Volcanic Activity during the Mid-Holocene," *Geology* 46 (2017): 47–50. One journalist for *The Atlantic* speculated about what would happen to Montserrat if another strong storm reached the island at the exact moment of a volcanic eruption. See Jim Nash, "What Would Happen if a Hurricane Hit an Erupting Volcano?," *The Atlantic*, October 18, 2017, https://www.theatlantic.com/science/archive/2017/10/hurricane-vs-volcano/543198/.

110. "Montserrat," Government of the United Kingdom.

111. "Climate Change," Organization of Eastern Caribbean States, accessed January 17, 2018, http://www.oecs.org/topics/climate-change; "Situation Analysis of Children in Montserrat," 26; "Montserrat Progresses with Its Development of a Climate Change Policy and Action Plan," Discover Montserrat, June 23, 2015, https://discovermni.com/2015/06/23/montserrat-progresses-with-its-development-of-a-climate-change-policy-and-action-plan/.

112. "Montserrat Progresses with Its Development of a Climate Change Policy and Action Plan."

113. IOM, "Data Collection 'Planned Relocation'—Montserrat," IOM Regional Office for Central America, North America and the Caribbean wesbite, accessed January 14, 2020, https://rosanjose.iom.int/SITE/en/vacancy/data-collection-planned-relocation-montserrat.

114. Eduardo A. Cavallo, Andrew Powell, and Oscar Becerra, "Estimating the Direct Economic Damage of the Earthquake in Haiti," Inter-American Development Bank, February 11, 2010, http://idbdocs.iadb.org/wsdocs/getdocument.aspx?docnum=35072649.

115. D'Vera Cohn, Jeffrey S. Passel, and Kristen Bialik, "Many Immigrants with Temporary Protected Status Face Uncertain Future in the U.S.," Pew Research Center, November 27, 2019, https://www.pewresearch.org/fact-tank/2019/11/27/immigrants-temporary-protected-status-in-us/.

Chapter 2

1. Hurricanes Cesar, Dolly, and Lili (1996), Mitch (1998), Keith and Gordon (2000), Iris and Michelle (2001), Rita, Beta, Adrian, Stan, and Wilma (2005), Felix and Dean (2007), Paloma (2008), Ida (2009), Richard, Alex, Karl, and Paula (2010), Rina (2011), Ernesto (2012), Barbara and Ingrid (2013), Sandra (2015), and Otto and Earl (2016).

2. These tropical storms are Kyle (1996), Andrés and Olaf (1997), Katrina (1999), Chantal (2001), Bill (2003), Adrian, Arlene, and Gamma (2005), Barry and Barbara (2007), Arthur and Alma (2008), Matthew and Agatha (2010), Harvey and Helene (2011), Barry (2013), and Boris and Hanna (2014). "Hurricanes and Tropical Cyclones," Weather Underground, accessed March 29, 2018, https://www.wunderground.com/hurricane/hurrarchive.asp?region=at.

3. "Database Search," Smithsonian Institution, National Museum of Natural History, Global Volcanism Project, accessed March 29, 2018, http://volcano.si.edu/search_eruption.cfm; William I. Rose, Gregg J. S. Bluth, Michael J. Carr, John W. Ewert, Lina C. Patino, and James W. Vallance, eds., *Volcanic Hazards in Central America* (Boulder, CO: Geological Society of America, 2006).

4. "ENSO and Drought Forecasting," University of Nebraska, National Drought Mitigation Center, accessed May 3, 2018, https://drought.unl.edu/Education/DroughtIn-depth/ENSO.aspx. The authors write, "Researchers have found the strongest connections between ENSO [El Niño–Southern Oscillation] and intense drought in Australia, India, Indonesia, the Philippines, Brazil, parts of east and south Africa, the western Pacific basin islands (including Hawaii), Central America, and various parts of the United States." See also "Disaster Risk Programme to Strengthen Resilience in the Dry Corridor of Central America," Food and Agriculture Organization of the United Nations, September 2015, http://www.fao.org/emergencies/resources/documents/resources-detail/en/c/330164/.

5. Juliana Martínez Franzoni and Diego Sánchez-Ancochea, *The Quest for Universal Social Policy in the South: Actors, Ideas and Architectures* (Cambridge: Cambridge University Press, 2016); Jason Hickel, "Want to Avert the Apocalypse? Take Lessons from Costa Rica," *The Guardian*, October 7, 2017, https://www.theguardian.com/working-in-development/2017/oct/07/how-to-avert-the-apocalypse-take-lessons-from-costa-rica.

6. *Food Insecurity and Emigration: Why People Flee and the Impact on Family Members Left Behind in El Salvador, Guatemala, and Honduras*, World Food Programme, August 2017, https://docs.wfp.org/api/documents/WFP-0000022124/download/.

7. Daniel Reichman, "Honduras: The Perils of Remittance Dependence and Clandestine Migration," *Migration Information Source*, April 11, 2013, http://www.migrationpolicy.org/article/honduras-perils-remittance-dependence-and-clandestine-migration.

NOTES TO PAGES 49-50

8. The great hurricane of 1780 killed 22,000 people in the eastern Caribbean. Neal Lott, Sam McCown, Axel Graumann, Tom Ross, and Mar Lackey, "Hurricane Mitch: The Deadliest Atlantic Hurricane since 1780," National Climatic Data Center, November 1999, ftp://ftp.ncdc.noaa.gov/pub/data/extremeevents/specialreports/Hurricane-Mitch-1998.pdf.

9. See as one example Gustavo Montenegro, "El monstro Mitch azotó el istmo hace 20 años," *Prensa libre* [Guatemala], October 28, 2018, https://www.prensalibre.com/guatemala/comunitario/el-monstruo-mitch-azoto-el-istmo-hace-20-aos/. Mark Schneider listed the casualties at 9,000. Subsequent tallies were much higher. "Text of Statement by the Honorable Mark L. Schneider, Assistant Administrator for Latin America and the Caribbean, U.S. Agency for International Development before the Subcommittee on Western Hemisphere Affairs, Committee on International Relations, U.S. House of Representatives, February 24, 1999," in file NSC [Immigration and Naturalization] [03/05/1999–03/31/1999], 2011-1044-F, box 1, Clinton Presidential Library (hereafter file NSC); Kathleen Newland, *Climate Change and Migration Dynamics*, Migration Policy Institute, September 2001, 7, http://www.migrationpolicy.org/research/climate-change-and-migration-dynamics; Oscar A. Ishizawa and Juan José Miranda, "Weathering Storms: Understanding the Impact of Natural Disasters on the Poor in Central America," Policy Research Working Paper No. 7692 (Washington, D.C.: World Bank, 2016), 2. Ishizawa and Miranda placed the number of casualties at 14,600.

10. "Hurricane Mitch: The United States Responds to Central America," February 16, 1999, in file "WHO [to Todd Stern] [02/15/1999–02/16/1999]," Automated Records Management System [email], 2013-0914-F (segment 2), box 13, Clinton Presidential Library.

11. "Text of Congressional Testimony of Brian Atwood, USAID Administrator," in email, Valerie N. Guarnieri to Wendy E. Gray et al., "USAID speeches on Mitch [unclassified]," March 2, 1999, NSC Emails Exchange-Non-Record (Mar 97–Jan 01) [Immigration and Naturalization] [05/13/1998–05/22/1998], NSC Records Management [Immigration and Naturalization] 95-2204, 2011-10440-F, box 9, Clinton Presidential Library (hereafter NSC Emails Exchange-Non-Record).

12. Press briefing, National Security Advisor Sandy Berger, Deputy Chief of Staff Maria Echavaste, USIA Administrator Brian Atwood, and Acting Assistant Secretary of State Peter Romero, March 4, 1999, in file "WHO [to Todd Stern] [03/04/1999] [1]," Automated Records Management System [email], 2013-0914-F, box 18, Clinton Presidential Library.

13. Bradley E. Ensor and Marisa Olivo Ensor, "Hurricane Mitch: Root Causes and Responses to Disaster," in *The Legacy of Hurricane Mitch: Lessons from Post-Disaster Reconstruction in Honduras*, ed. Marisa O. Ensor (Tucson: University of Arizona Press, 2009), 22.

14. *Central America: Assessment of the Damage Caused by Hurricane Mitch, 1998: Implications for Economic and Social Development and for the Environment*, United Nations Economic Commission for Latin America and the Caribbean, May 21, 1999, 8–9, http://repositorio.cepal.org//handle/11362/25513.

15. *Nicaragua: Assessment of the Damage Caused by Hurricane Mitch, 1998: Implications for Economic and Social Development and for the Environment*, United Nations Economic Commission for Latin America and the Caribbean, April 19, 1999, 7, https://www.cepal.org/publicaciones/xml/7/15507/L372-1-EN.pdf.

16. Glenn Garvin, "Mitch Death Toll Tops 1,500, Thousands Missing," *Miami Herald*, November 2, 1998, 1A.

17. Estelí, Nueva Segovia, Madriz, Managua, Granada, Masaya, Carazo, Boaco, Matagalpa, Jinotega, Waspan and Cruz Rio Grande, and the Región Autónoma del Atlántico Norte (RAAN) were also badly affected. *Nicaragua: Assessment of the Damage Caused by Hurricane Mitch, 1998*; Sarah Stewart, "Nicaragua: Reflexiones sobre huracán Mitch y sus secuelas," Fundación Inter-americana, December 16, 2018, https://www.iaf.gov/es/content/publicacion/nicaragua-reflexiones-sobre-el-huracan-mitch-y-sus-secuelas/; Garvin, "Mitch Death Toll Tops 1,500"; Steven Thomma, "Clinton Brings Aid Plan to Nicaragua President; Sees Devastation Left by Mitch," *Miami Herald*, March 9, 1999, 3A.

18. The storm severely damaged 94,000 buildings, including 512 schools and 250 health care facilities. The cost of providing new housing and furnishings for those displaced by the storm was estimated at $425 million. The storm also destroyed or damaged seventy-one bridges and rendered 8,300 miles of roads impassable. *Nicaragua: Assessment of the Damage Caused by Hurricane Mitch, 1998*, 33–36.

19. Julie Watson, "Mitch Leaves Scattered Mines," *Miami Herald*, December 3, 1998, 16A; *Nicaragua: Assessment of the Damage Caused by Hurricane Mitch, 1998*, 15; "Many Victims Wait in Vain for Help to Arrive," *Miami Herald*, November 7, 1998, 5A; Arnold Markowitz, "Mitch Wet, Mighty at 180 MHP," *Miami Herald*, October 27, 1998, 1A; Glenn Garvin and Juan O. Tamayo, "Crisis in Central America," *Miami Herald*, November 22, 1998, 1HM.

20. See Elizabeth Beery Adams, "The History of Demining in Nicaragua," *Journal of Conventional Weapons Destruction* 5 (August 2001): 34–37, http://commons.lib.jmu.edu/cgi/viewcontent.cgi?article=2199&context=cisr-journal; Watson, "Mitch Leaves Scattered Mines."

21. Psychologist Josefina Murillo Vargas reflected on Mitch's human impacts on Nicaragua and collected testimonies and newspaper clippings in *Testimonios de amor y esperanza: Damnificados del Huracán Mitch* (Managua: Ediciones Graphic Prints, 1999). This particular quote appears on page 28 (translation mine). See also the accounts of the journalists working for the *Miami Herald*: Glenn Garvin, "Youngest Survivors of Hurricane Mitch Get Special Treatment," November 22, 1998, 1A; Glenn Garvin, "After Mitch: A Year after Hurricane, Devastated Lives Are Healing—Some More Speedily Than Others," November 7, 1999, 1L; Glenn Garvin and Arnold Markowitz, "Mitch: The Aftermath Ruin and Despair—the Toll Climbs in Central America; Known Deaths Climb to 1,700," November 3, 1998, 1A; Glenn Garvin and Arnold Markowitz, "Mitch Pours It On," November 1, 1998, 1A; Andres Viglucci, "Central America Paralyzed by Mitch," November 8, 1998, 1A; "Central America Gets More Help; Hillary Rodham Clinton Announces Doubling of U.S. Aid," November 17, 1998, 24A; Arnold Markowitz, "A Savage Season in the Tropics Reaches End," November 30, 1998, 1A.

22. Garvin, "Youngest Survivors of Hurricane Mitch Get Special Treatment."

23. *Central America: Assessment of the Damage Caused by Hurricane Mitch, 1998*, 26; *Nicaragua: Assessment of the Damage Caused by Hurricane Mitch, 1998*, 33, 47; Oxfam America, Christian Aid, and Universidad Centroamericana, *Mitch y desarrollo rural: El impacto del huracán Mitch en Nicaragua y sus implicaciones para las prioridades de desarrollo sostenible de las ONGs y organizaciones sociales* (Nitlapán, Nicaragua: Universidad Centroamericana, 1999); U.S. Department of Agriculture, "Nicaragua and Hurricane Mitch: 2nd Preliminary Report, 1998," in file "Global Agriculture Information Network (GAIN) Reports, 1998," Global Agriculture Information Network (GAIN) Reports, 7/1/1998–12/31/2011, RG166, Records of the Foreign Agricultural Service, National Archives, College Park, Maryland.

24. The value of lost exports amounted to $54 million, or 7 to 8 percent of foreign sales. *Nicaragua: Assessment of the Damage Caused by Hurricane Mitch, 1998*, 45.

25. "Remarks by the First Lady and Mrs. Gore," February 16, 1999, in File WHO [To: Todd Stern] in Clinton Presidential Records Automated Records Management System [Email] 2013-0914-F (Segment 2), Box 14 of 18.

26. *Central America: Assessment of the Damage Caused by Hurricane Mitch, 1998*, 9.

27. *Honduras: Assessment of the Damage Caused by Hurricane Mitch, 1998: Implications for Economic and Social Development and for the Environment*, United Nations Economic Commission for Latin America and the Caribbean, April 14, 1999, 8, https://www.cepal.org/publicaciones/xml/6/15506/L367-1-EN.pdf.

28. Manuel Torres, *Huracán Mitch 1998–2003: Retrato social de una tragedia natural* (Tegucigalpa: Centro de Documentación de Honduras, 2004), 41 (translation mine).

29. Glenn Garvin and Arnold Markowitz, "Flooding in Honduras Hides Storm's Death Toll," *Miami Herald*, October 30, 1998, 1A.

30. Michael W. W. Ottey, "Hurricane Mitch Still Scars Honduras," *Miami Herald*, November 5, 2003, 12A; Ensor and Ensor, "Hurricane Mitch," 40.

31. Garvin, "After Mitch."

32. Thelma Mejía, "La reforma política y la participación de sociedad civil," in *Descifrando a Honduras: Cuatro puntos de vista sobre la realidad política tras el huracán Mitch*, eds. Manuel Torres Calderón, Thelma Mejía, Dan Alder, and Paul Jeffrey (Cambridge, Mass.: Hemisphere Initiatives, August 2002), 24–33; Ensor and Ensor, "Hurricane Mitch," 37.

33. Ensor and Ensor, "Hurricane Mitch," 38; Edwin L. Harp, Mario Castañeda, and Matthew D. Held, *Landslides Triggered by Hurricane Mitch in Tegucigalpa, Honduras* (Washington, D.C.: U.S. Department of the Interior, U.S. Geological Survey, n.d.), 1, accessed January 19, 2021, https://pubs.usgs.gov/of/2002/ofr-02-0033/OFR0233_Text.access.pdf.

34. Garvin, "After Mitch."

35. Garvin and Markowitz, "Mitch: The Aftermath Ruin and Despair."

36. Cited in Sarah Stewart, "Honduras: Reflexiones sobre el huracán Mitch y sus secuelas," Fundación Inter-americana, December 16, 2018, https://www.iaf.gov/es/content/publicacion/honduras-reflexiones-sobre-el-huracan-mitch-y-sus-secuelas/.

37. *Honduras: Assessment of the Damage Caused by Hurricane Mitch, 1998*, 10; Madeline Messick and Claire Bergeron, "Temporary Protected Status in the United States: A Grant of Humanitarian Relief That Is Less Than Permanent," *Migration Information*

Source, July 2, 2014, http://www.migrationpolicy.org/article/temporary-protected-status-united-states-grant-humanitarian-relief-less-permanent; Garvin, "Mitch Death Toll Tops 1,500."

38. "Text of Oral Statement to Congress by Mark Schneider, USAID Assistant Administrator for Latin America," in email, Valerie N. Guarnieri to Wendy E. Gray et al., "USAID Speeches on Mitch [unclassified]," March 2, 1999; Messick and Bergeron, "Temporary Protected Status in the United States."

39. Anthony Oliver-Smith, "Disasters and Forced Migrations in the 21st Century," Social Science Research Council, June 11, 2006, http://understandingkatrina.ssrc.org/Oliver-Smith/.

40. Joint Global Change Research Institute, Battelle Memorial Institute, Pacific Northwest Division, and National Intelligence Council, *Mexico, the Caribbean, and Central America: The Impact of Climate Change to 2030: A Commissioned Research Report*, December 2009, 24–25, http://globaltrends.thedialogue.org/publication/mexico-the-caribbean-and-central-america-the-impact-of-climate-change-to-2030/.

41. Text of oral statement to Congress by Mark Schneider, March 2, 1999.

42. Watson, "Mitch Leaves Scattered Mines," 16A. On Friday, October 8, 2004, the government of Honduras convened a ceremony on the outskirts of Tegucigalpa to officially declare Honduras mine-free. The United States and other nations, working through the Organization of American States and the Inter-American Defense Board, began work on mine clearance in 1994 and continued for a total of nine years. The team was responsible for clearing an area that totaled 1,479 square kilometers and resulted in the removal of approximately 2,191 mines, 214 explosive artifacts, and 60,521 pieces of war refuse. According to one U.S. embassy telegram, "With the land now clear of danger, the [government of Honduras] estimates that more than 67,000 families will likely move into the area and start farming." Inter-American Defense Board personnel reported, however, that due to Hurricane Mitch, "there may be some deep-seated mines that have yet to be discovered. These mines would only pose a potential problem if the area were to undergo major construction projects, which could disturb and perhaps set off the mines." Telegram (cable) from Ambassador Larry Leon Palmer, "Honduras Declared Mine Free," WikiLeaks, October 22, 2004, https://wikileaks.org/plusd/cables/04TEGUCIGALPA2373_a.html.

43. U.S. Department of Agriculture, "Honduras: Agricultural Situation, Hurricane Mitch Update, 1998," in file "Global Agriculture Information Network (GAIN) Reports, 1998," Global Agriculture Information Network (GAIN) Reports, 7/1/1998–12/31/2011, RG166, Records of the Foreign Agricultural Service, National Archives; Malcolm Rodgers, "In Debt to Disaster: What Happened to Honduras after Hurricane Mitch," Christian Aid, October 31, 1999, https://reliefweb.int/report/honduras/debt-disaster-what-happened-honduras-after-hurricane-mitch.

44. *Honduras: Assessment of the Damage Caused by Hurricane Mitch, 1998*, 55; Glenn Garvin and Jane Bussey, "Loss of Crops, Roads Cripples Hondurans," *Miami Herald*, November 9, 1998, 1A. Over sixty highways and ninety bridges were damaged. Shrimp farms suffered an estimated $100 million in damages.

45. Garvin and Markowitz, "Mitch: The Aftermath Ruin and Despair."

46. Edward Hegstrom, "Reporting on Mitch Overblown," *Pittsburgh Post-Gazette*, December 13, 1998, Nexis Uni; Julie Watson, "Honduras Lowers Mitch Death Toll," Associated Press, December 1, 1998, https://apnews.com/article/5eb87c118a1f27b8b18eeae0755644b9.

47. Glenn Garvin and Arnold Markowitz, "Catástrofe en Centroamérica—miles aún esperan por rescate," *El Nuevo Herald*, November 3, 1998, 1A (translation mine); Lycia Naff, "Faith in Future Key in Honduras, Bishop Says," *Miami Herald*, December 28, 1998, 1B.

48. William M. Loker, "A Flood of Impressions: Riding Out Mitch and Its Aftermath," in Ensor, *Legacy of Hurricane Mitch*, 97; see also Ensor and Ensor, "Hurricane Mitch," 40.

49. Cited in Stewart, "Honduras: Reflexiones sobre el huracán Mitch y sus secuelas."

50. Mejía, "La reforma política y la participación de sociedad civil," 24–33.

51. Rodgers, "In Debt to Disaster."

52. Paul Jeffrey, "Central America Turns to Churches for Help after Hurricane Mitch," *Ecumenical News International Bulletin*, December 2, 1998, 7, cited in Rodgers, "In Debt to Disaster."

53. Telegram (cable) from Ambassador Larry Leon Palmer, "Honduran Fiscal Transparency and Accountability," WikiLeaks, May 5, 2003, https://wikileaks.org/plusd/cables/03TEGUCIGALPA1041_a.html.

54. Economic and Social Council, "Report of the Secretary General: Collaborative Efforts to Assist Belize, Costa Rica, El Salvador, Guatemala, Honduras, Nicaragua and Panama and Progress Made with the Relief, Rehabilitation and Reconstruction Efforts of the Affected Countries," United Nations, July 5–30, 1999, https://www.un.org/esa/documents/ecosoc/docs/1999/e1999-72.htm.

55. According to U.S. embassy officials, Taiwan had a long history of assistance to Nicaragua. It was one of the first countries to deliver assistance after Hurricane Mitch, and Taiwan built Nicaragua's foreign ministry, former presidential offices, and national assembly building in Managua. Telegram (cable) from Ambassador Paul A. Trivelli, "Checkbook Diplomacy? Taiwanese President Visits Nicaragua," WikiLeaks, September 21, 2007, https://wikileaks.org/plusd/cables/07MANAGUA2169_a.html.

Among the international relief workers were 200 Cuban doctors sent by the Castro government. See John M. Kirk and H. Michael Erisman, *Cuban Medical Internationalism: Origins, Evolution, and Goals* (New York: Palgrave Macmillan, 2009). By 2007, there were approximately 280 Cuban medical doctors working in Honduras. In October 1999, Honduras and Cuba signed an "Agreement on Health Cooperation," allowing Cuban medical brigades to provide health care in the most underdeveloped areas of the country, financed by the government of Honduras. Cuba also provided scholarships for Hondurans to study medicine at the Escuela Latinoamericana de Medicina in Havana. Approximately 300 Honduran students began the six-year program at the school in 1999, and 218 graduated in 2005. In 2006 the Honduran press reported that the mayor of San Pedro Sula and the Cuban ambassador to Honduras signed an agreement to arrange for the Cuban government's provision of free reading and writing lessons in the municipality. See telegram (cable) from U.S. embassy in Tegucigalpa, "Honduras:

Review for Suspension of Title III of the Libertad Act," WikiLeaks, November 28, 2007, https://wikileaks.org/plusd/cables/07TEGUCIGALPA1836_a.html.

56. Vilma Elisa Fuentes, "Post-Disaster Reconstruction: An Opportunity for Political Change," in Ensor, *Legacy of Hurricane Mitch*, 106. See also Ensor and Ensor, "Hurricane Mitch," 39; and Rodgers, "In Debt to Disaster."

57. Press briefing, Berger, Echavaste, Atwood, and Romero, March 4, 1999.

58. Text of oral statement to Congress by Mark Schneider, March 2, 1999.

59. Ernesto Paz, "The Foreign Policy and National Security of Honduras," in *Honduras Confronts Its Future: Contending Perspectives on Critical Issues*, ed. Mark B. Rosenberg and Philip L. Shepherd (Boulder: Rienner, 1986), cited in Ensor and Ensor, "Hurricane Mitch," 30.

60. For a discussion of U.S. policy in Central America, see Walter F. LaFeber, *Inevitable Revolutions: The United States in Central America* (New York: W. W. Norton, 1984).

61. Letter, Honorable Raul Felipe Torres, Consul General of Honduras, to President William J. Clinton, September 24, 1997, in file [Free Trade Area] 9706692, NSC Records Management, box 2017-0397-F [Free Trade Area 9409314 to 0004038], Clinton Presidential Library.

62. "Remarks by the First Lady and Mrs. Gore," February 16, 1999.

63. Soto Cano is home to the Joint Task Force Bravo of the U.S. Southern Command. The U.S. Southern Command area of responsibility encompasses thirty-one countries and ten territories. "Joint Task Force-Bravo," U.S. Department of Defense, accessed May 14, 2018, http://www.jtfb.southcom.mil/About-Us/. During the 1980s, Soto Cano served as the staging ground for U.S. support for the Contras in neighboring Nicaragua. In the 1990s, when the U.S. Southern Command's mission moved from fighting communism to the "War on Drugs," its budget increased more than any other U.S. military regional command, and Soto Cano became critical to that mission. Jake Johnston, "How Pentagon Officials May Have Encouraged a 2009 Coup in Honduras," *The Intercept*, August 29, 2017, https://theintercept.com/2017/08/29/honduras-coup-us-defense-departmetnt-center-hemispheric-defense-studies-chds/.

64. In Nicaragua alone, USAID delivered "679 rolls of plastic sheeting . . . [for] more than 2,700 shelters. A total of 15,000 water jugs and 14,000 blankets[,] . . . rain gear and boots, cooking utensils, medicine, safety equipment, flashlights, latrines, home water purification systems, first aid kits, and 50,000 mosquito nets" were also delivered. See "Text of Statement by the Honorable Mark L. Schneider," February 24, 1999.

65. "Text of Statement by the Honorable Mark L. Schneider," February 24, 1999.

66. Gore was accompanied by Brian Atwood of USAID, Nicaraguan baseball pitcher Dennis Martínez of the Atlanta Braves, and John Leonard, deputy assistant secretary of state for Latin America. The congressional delegation included Democratic senators Christopher Dodd of Connecticut, Jeff Bingaman of New Mexico, and Mary Landrieu of Louisiana; Republican senator Jim Colby of Arizona; and Democratic representatives Gary Ackerman of New York and Xavier Becerra of California. Karen Testa, "Tipper Gore Visits Honduras," Associated Press, November 11, 1999, https://www.apnews.com/d9c21c6a81a6a5d29bb77c7e8c3e9ad8; "Tipper Gore Visits Nicaragua and Honduras in Wake of Hurricane Mitch," National Archives, accessed May 21, 2018, https://

clintonwhitehouse3.archives.gov/WH/EOP/VP_Wife/trips/hurricane.html; "Remarks by the First Lady and Mrs. Gore," February 16, 1999.

67. First Lady's Office, Press Office, and Lissa Muscatine, "FLOTUS Statements and Speeches 6/27/98–11/30/98 [Binder]: [Arrival in Managua, Nicaragua 11/16/98]," Clinton Digital Library, accessed February 3, 2017, https://clinton.presidentiallibraries.us/items/show/8358. USAID increased emergency food aid by an additional $15 million. This brought the total food aid to $25 million by the end of November. Guatemala and El Salvador were offered $10 million in concessional loans for food purchases.

68. "Text of Congressional Testimony of Brian Atwood," March 2, 1999.

69. "The Denton Program for Private Donations," USAID, accessed March 10, 2018, https://www.usaid.gov/work-usaid/partnership-opportunities/humanitarian-responders/denton-program.

70. They had repaired 700 damaged health clinics and 1,700 schools; provided clean water and sanitation services for 7 million people; repaired 700 kilometers of rural roads; provided school supplies for 200,000 children; built 6,400 new housing units; provided farm supplies to 5 million people; and trained mayors and staff from over 100 severely damaged cities and towns to plan and manage resources. Email, Scott W. Busby to Richard L. Denniston, "Gobush Q&As with Immigration Changes [unclassified]," February 16, 1999, in file "Exchange Record (Sept 97–Jan 01)," 2011-10440-F, box 6, Clinton Presidential Library.

71. Press briefing, Berger, Echavaste, Atwood, and Romero, March 4, 1999. See also Office of the Press Secretary, "Statement by the Press Secretary, First Lady Hillary Rodham Clinton Announces Major Expansion of U.S. Relief Effort in Central America," November 16, 1998, in file "Exchange Record (Sept 97–Jan 01) [06/11/1998–12/28/1998]," 2011-10440-F, box 6, Clinton Presidential Library.

72. The earthquake on January 26, 1999, measuring 6.0 on the Richter scale, struck west of Bogotá, killing 1,000 people, injuring 3,500, and leaving 35,000 people homeless.

73. Email, Scott W. Busby to Richard L. Denniston, "Gobush Q&As with Immigration Changes [unclassified]," February 16, 1999; press briefing, Berger, Echavaste, Atwood, and Romero, March 4, 1999.

74. Email, Natalie S. Wozniak to Natalie S. Wozniak, "Finish—$1 Billion [unclassified]," February 16, 1999, Clinton Presidential Records, NSC Emails Exchange-Non-Record.

75. Email, Valerie N. Guarnieri to Wendy E. Gray et al., "USAID speeches on Mitch [unclassified]," March 2, 1999.

76. "Remarks by the First Lady and Mrs. Gore," February 16, 1999.

77. The Paris Club has nineteen permanent members, including most of the Western European and Scandinavian nations, the United States, the United Kingdom, and Japan.

78. Fuentes, "Post Disaster Reconstruction," 108.

79. Fuentes, 108. Manuel Torres discusses the role of civil society organizations in the chapter "El Post Mitch de la sociedad civil" in *Huracán Mitch 1998–2003*, 77–88. Rodgers also discusses the role of civil society organizations in "In Debt to Disaster."

80. Email, Valerie N. Guarnieri to Wendy E. Gray et al., "USAID speeches on Mitch [unclassified]," March 2, 1999; email, Valerie N. Guarnieri to Philip C. Bobbitt et al.,

"FW: Gobush Q&As with immigrant changes [unclassified]," February 6, 1999, in file "Exchange Record (Sept 97–Jan 01) [02/03/1999–/2/28/1999]," 2011-10440-F, box 6, Clinton Presidential Library.

81. Ensor and Ensor, "Hurricane Mitch," 30; Carmen Reinhart and Cristoph Trebesh, "The International Monetary Fund: 70 Years of Reinvention," National Bureau of Economic Research, Working Paper 21805 (December 2015), doi 10.3386/w21805; "1998 Corruption Perceptions Index," Transparency International, accessed January 19, 2021, https://www.transparency.org/en/press/1998-corruption-perceptions-index.

82. Rodgers, "In Debt to Disaster." Rodgers discusses the "Paris moratorium" that proposed an emergency mechanism to freeze debt repayments by countries hit hard by natural disasters.

83. Email, Scott W. Busby to Richard L. Denniston, "Gobush Q&As with Immigration changes [unclassified]," February 16, 1999. See also press briefing, Berger, Echavaste, Atwood, and Romero, March 4, 1999; and "Hurricane Mitch: The United States Responds to Central America," February 16, 1999, in file "WHO [to Todd Stern] [02/15/1999–02/16/1999]," Automated Records Management System [email], 2013-0914-F (segment 2), box 13, Clinton Presidential Library. See also email, Natalie S. Wozniak to Natalie S. Wozniak, "Finish—$1 Billion [unclassified]," February 16, 1999.

84. "Hurricane Mitch: The United States Responds to Central America."

85. "Central American Debt: Paris Club 'Just a Stay of Execution,' Says Oxfam," *Oxfam*, December 10, 1998, https://reliefweb.int/report/honduras/central-american-debt-paris-club-just-stay-execution-say-oxfam; Rodgers, "In Debt to Disaster."

86. Office of the Press Secretary, "Statement by the Press Secretary, First Lady Hillary Rodham Clinton Announces Major Expansion of U.S. Relief Effort in Central America."

87. Ensor and Ensor, "Hurricane Mitch," 34. David Bacon wrote that by the early 1990s, women made up 84 percent of the workforce in Honduran maquiladoras, and over half were younger than twenty (some as young as ten). To keep women from getting pregnant and leaving the factory to have children, USAID funded the Honduran Association for Family Planning, which established "contraceptive distribution posts" staffed by nurses in three factories. David Bacon, "If San Pedro Sula Is the Murder Capital of the World, Who Made It That Way?," *American Prospect*, June 13, 2019, https://prospect.org/economy/san-pedro-sula-murder-capital-world-made-way/.

88. The Caribbean Basin Initiative was established by the 1983 Caribbean Basin Economic Recovery Act and expanded by the Caribbean Basin Economic Recovery Expansion Act of 1990. The countries did not achieve "NAFTA-parity" until the 2000 Caribbean Basin Trade Partnership Act.

89. The proposal provided duty-free and quota-free treatment of textile and apparel products if they were one of three categories: (1) assembled in the region of fabric made in the United States from U.S. yarn and cut in the United States; (2) cut and assembled in the region from U.S. fabric made from U.S. yarn; or (3) handmade, folklore, hand-loomed articles. See press briefing, Berger, Echavaste, Atwood, and Romero, March 4, 1999.

90. Letter, Jeffrey De Laurentis, Director, Inter-American Affairs, to the Honorable Raul Felipe Torres, Consul General of Honduras, November 10, 1997, in file [Free Trade

Area] 9706692, NSC Records Management, 2017-0397-F [Free Trade Area 9409314 to 0004038], box 1, Clinton Presidential Library. See also email, Nisha Desai to Valerie N. Guarnieri, "Text of Testimony of J. Brian Atwood, February 24, 1999," March 5, 1999, in file NSC.

91. Ministry of Foreign Affairs of the Republic of Honduras, "Request for Extension of Designation of Honduras under Temporary Protected Status (TPS) Program," March 1, 2000, 8, Clinton Digital Library.

92. Email, Nisha Desai to Valerie N. Guarnieri, "Mark Schneider Oral," March 5, 1999, in file NSC.

93. Email, Valerie N. Guarnieri to Wendy E. Gray et al., "USAID speeches on Mitch [unclassified]," March 2, 1999. See also "Text of Statement by the Honorable Mark L. Schneider," February 24, 1999.

94. Email, Nisha Desai to Valerie N. Guarnieri, "Text of Testimony of J. Brian Atwood, February 24, 1999," March 5, 1999, in file NSC.

95. Email, Nisha Desai to Valerie N. Guarnieri, "Mark Schneider Oral," March 5, 1999; "Text of Statement by the Honorable Mark L. Schneider," February 24, 1999.

96. Press briefing, Berger, Echavaste, Atwood, and Romero, March 4, 1999.

97. In one of his speeches, Clinton remarked, "This is the first time a President of the United States has been anywhere in Guatemala outside of the airport, which President Johnson visited briefly more than thirty years ago. It is a visit long overdue." National Security Council, Speechwriting Office, and Edward (Ted) Widmer, "Guatemala [March 11, 1999]," Clinton Digital Library, https://clinton.presidentiallibraries.us/items/show/11466.

98. The summit was attended by the presidents of Costa Rica, El Salvador, Guatemala, Honduras, Nicaragua, the Dominican Republic, and Belize. "Declaración presidencial de Antigua Guatemala," in Torres, *Huracán Mitch 1998–2003*, 118–27 (translation mine).

99. James Gerstenzang and Juanita Darling, "Clinton Gives Apology for U.S. Role in Guatemala," *Los Angeles Times*, March 11, 1999, https://www.latimes.com/archives/la-xpm-1999-mar-11-mn-16261-story.html.

100. Much has been written about this episode. See, for example, Daniel Beckman, "A Labyrinth of Deception: Secretary Clinton and the Honduran Coup," Council on Hemispheric Affairs, April 12, 2017, https://www.coha.org/a-labyrinth-of-deception-secretary-clinton-and-the-honduran-coup/. Zelaya had announced a program of economic and social reforms, including raising the minimum wage, giving subsidies to small farmers, cutting interest rates, and instituting free education, but he was opposed by powerful growers and corporate mining interests, among them Miguel Facussé, the billionaire uncle of former Honduran president Carlos Flores Facussé.

101. Norwegian Agency for Development Cooperation, *Joint Evaluation of Support to Anti-corruption Efforts, Nicaragua Country Report*, June 2011, 14, https://www.oecd.org/countries/tanzania/48912912.pdf; Jenny Pearce and Andrew Bounds, "Natural and Unnatural Disasters in Central America," in *South America, Central America, and the Caribbean 2003*, ed. Jacqueline West (London: Europa Group, 2002), 41. The U.S. embassy in Managua reported that in the spring of 1999, while Nicaraguans were recovering from the hurricane's devastating consequences, President Arnoldo Alemán

took his entire extended family with him to Sweden "so that they could enjoy a vacation and go shopping at government expense" while he attended a hurricane reconstruction meeting. "On this and other trips, the Alemán family ran up a combined total of nearly $2 million in personal expenses on government credit cards, and had Central Bank President Noel Ramírez write checks to make the Nicaraguan people pay for the Alemán family's private shopping and travels. Documentation on all of this spending is in the hands of the Nicaraguan and United States governments." Telegram from Ambassador Paul A. Trivelli, "Nicaragua's Most Wanted Part III: The Crimes of Arnoldo Alemán and his Family," WikiLeaks, May 5, 2006, https://wikileaks.org/plusd/cables/06MANAGUA1004_a.html.

102. Manuel Torres Calderón, "Quién conoce Honduras?," in Torres Calderón et al., *Descifrando a Honduras*, 12; Paul Jeffrey, "Invasión Campesina a una Antigua Base Militar," in Torres Calderón et al., *Descifrando a Honduras*, 56; "Honduras: Human Rights Violations in Bajo Aguán," Business and Human Rights Resource Centre, July 2011, https://www.europarl.europa.eu/meetdocs/2009_2014/documents/dcam/dv/8_ffm_report_bajo_aguan_/8_ffm_report_bajo_aguan_en.pdf; Rodgers, "In Debt to Disaster."

103. William Branigin and Roberto Suro, "U.S. Seeks to Stem a Wave of Migration after Mitch; Relief Plan Would Touch Thousands in D.C. Area," *Washington Post*, November 14, 1998, A1, https://www.washingtonpost.com/archive/politics/1998/11/14/us-seeks-to-stem-a-wave-of-migration-after-mitch/dc477130-a97f-494d-a1ca-22f8da2b07e9/.

104. The Agricultural Modernization Law of 1992, in particular, had undermined the agrarian reform agenda of the previous two decades. Torres Calderón, "Quién conoce Honduras?," 11–12.

105. Maria Cristina Garcia, *Seeking Refuge: Central American Migration to Mexico, the United States, and Canada* (Berkeley: University of California Press, 2006).

106. Most of the applicants were the immediate relatives of U.S. citizens and permanent residents of Central American ancestry. Dean Yang and Parag Mahajan, "Hurricanes Drive Immigration to the United States," *The Conversation*, September 15, 2017, https://theconversation.com/hurricanes-drive-immigration-to-the-us-83755.

107. Two other amnesty programs were announced during the 1990–94 period. The State of the Nation Project, *A Binational Study: The State of Migration Flows between Costa Rica and Nicaragua*, International Organization for Migration, December 2001, 44, https://publications.iom.int/books/binational-study-state-migration-flows-between-costa-rica-and-nicaragua. See also "A Silver Lining," *The Economist*, August 12, 1999, https://www.economist.com/the-americas/1999/08/12/a-silver-lining.

108. CA-4, the Central America 4 Border Control Agreement between Nicaragua, El Salvador, Honduras, and Guatemala, established free movement for its citizens across their borders. Visitors who enter one of the four countries can travel to the other three without acquiring an additional visa. However, visitors are allowed to remain only ninety days total in these countries and are required to file for an extension if they wish to remain beyond ninety days. See "Immigration Laws," U.S. Embassy in Nicaragua, accessed August 8, 2018, https://ni.usembassy.gov/u-s-citizen-services/citizenship-services/immigration-laws/; and "Mitch Leads to TPS," *Migration News*, January 1999, https://migration.ucdavis.edu/mn/more.php?id=1689.

109. Garcia, *Seeking Refuge*, 158–61.

110. Email, Robert Malley to Samuel Berger and James Steinberg, "Immigration-background/tps/qs & as [unclassified]," April 30, 1997, in file MSMail-Record (Sept 94–Sept 97) [04/30/1997], NSC Emails, 2011-10440-F, box 5, Clinton Presidential Library.

111. Ricardo Sandoval, "In Wake of Mitch, Massive Illegal Immigration Is Expected," *Miami Herald*, April 17, 1999, 4A; Andres Viglucci, "Influx of Refugees after Hurricane Mitch Now Seems Unlikely," *Miami Herald*, August 6, 1999, 1B.

112. Ginger Thompson, "Storm Victims Surge North, with U.S. as Goal," *New York Times*, January 18, 1999, A1.

113. Email, Valerie N. Guarnieri to Wendy E. Gray et al., "USAID speeches on Mitch [unclassified]," March 2, 1999.

114. Press briefing, Berger, Echavaste, Atwood, and Romero, March 4, 1999.

115. Maria Cristina Garcia, "Refuge in the National Security State," in *The Refugee Challenge in Post–Cold War America* (New York: Oxford University Press, 2017), 113–57. See also Maria Cristina Garcia, "National (In)security and the 1996 Immigration Act," *Modern American History* 1, no. 2 (July 2018): 233–36.

116. Email, Robert Malley to William C. Danvers, "Migration Points [unclassified]," April 30, 1997, in file MSMail-Record (Sept 94–Sept 97) [04/30/1997], 2011-10440-F, box 5, Clinton Presidential Library.

117. [Name redacted], "The Nicaraguan Adjustment and Central American Relief Act: Hardship Relief and Long-Term Illegal Aliens," Congressional Research Service, July 15, 1998, 4, https://www.everycrsreport.com/files/19980715_98-3_08ea932f-fbb5b70b21888bb84863bfba90bfba25.pdf; "Suspension of Deportation and Cancellation of Removal," *Federal Register* 62 (September 30, 1998): 52134–40, https://www.federalregister.gov/documents/1998/09/30/98-26200/suspension-of-deportation-and-cancellation-of-removal; Carl Shusterman, "BIA Defines Hardship Standard for Cancellation of Removal," *Immigration Daily*, accessed August 4, 2018, https://www.ilw.com/articles/2001,0614-Shusterman.shtm.

118. Email, Robert Malley to Samuel Berger and James Steinberg, "Immigration-background/tps/qs & as [unclassified]," April 30, 1997.

119. The case was initially known as *American Baptist Churches of the USA v. Meese*. "American Baptist Churches v. Thornburgh (ABC) Settlement Agreement," USCIS, accessed August 16, 2021, https://www.uscis.gov/humanitarian/refugees-and-asylum/asylum/american-baptist-churches-v-thornburgh-abc-settlement-agreement.

120. During the 1980s, fewer than 5 percent of Salvadorans and Guatemalans received asylum. Nicaraguans were comparatively more successful. See Garcia, *Seeking Refuge*, chapter 2.

121. Letter, Howard Berman and seventeen other congressmen to Doris Meissner, INS Commissioner, March 5, 1997, in file [Immigration and Naturalization], NSC Emails Exchange-Non-Record.

122. Memorandum, Samuel Berger and John Hilley to President William J. Clinton, "Response to Representative McGovern Letter on Immigration Bill," April 30, 1997, in file [Immigration and Naturalization], 2011-10440-F, box 9, Clinton Presidential Library; see also email, Robert Malley to William C. Danvers, "Migration Points [unclassified]," April 30, 1997.

123. Email, Robert Malley to William C. Danvers, "Migration Points [unclassified]," April 30, 1997; email, Robert Malley to Samuel Berger and James Steinberg, "Immigration-background/tps/qs & as [unclassified]," April 30, 1997.

124. Letter, Jeffrey De Laurentis to Raul Felipe Torres, November 10, 1997; letter, Raul Felipe Torres to William J. Clinton, September 24, 1997. See also "Central America: Parity Legislation Options Outline (INS Draft 3/31/99)," in file Mandatory Detention, WHORM Subject File-General FG006-01 to Domestic Policy Council, 2011-1044-F, box 1, Clinton Presidential Library.

125. [Name redacted], "Nicaraguan Adjustment and Central American Relief Act"; Executive Office for Immigration Review, "Procedures Further Implementing the Annual Limitation on Suspension of Deportation and Cancellation of Removal," *Federal Register* 82 (December 5, 2017): 57336–40, https://www.federalregister.gov/documents/2017/12/05/2017-26104/procedures-further-implementing-the-annual-limitation-on-suspension-of-deportation-and-cancellation.

126. [Name redacted], "Nicaraguan Adjustment and Central American Relief Act"; memorandum, Anthony Lake to President William J. Clinton, July 17, 1996, in file MSMail-Record (Sept 94–Sept 97) [06/04/1996–12/18/1996], 2011-10440-F, box 5, Clinton Presidential Library.

127. The lawsuit had the support of Rep. Lincoln Díaz-Balart of Florida, one of the legislative architects of NACARA, as well as Reps. Ileana Ros-Lehtinen of Florida and Bob Livingston of New Orleans. Email, Jeffrey De Laurentis to Scott W. Busby, "Honduras-USA-IMIGRAT [unclassified]," February 4, 1998, in file MSMail-Record (Sept 94–Sept 97), 2011-1044-F, box 6, Clinton Presidential Library.

128. "Central America: Parity Legislation Options Outline (INS Draft 3/31/99)."

129. Email, Jeffrey De Laurentis to Scott Busby, "Honduras-USA-Immigrat [unclassified]," February 4, 1998, in file "MSMail-Non-record (Sept 94–Sept 97) [02/06/1997–05/13/1997], 2011-10440-F, box 6, Clinton Presidential Library.

130. Sarah Blanchard, Erin Hamilton, Nestor Rodriguez, and Hirotoshi Yoshioka, "Shifting Trends in Central American Immigration: A Demographic Examination of Increasing Honduran-U.S. Immigration and Deportation," *Latin Americanist* 55, no.4 (December 2011), 61–84, https://liberalarts.utexas.edu/etag/_files/pdfs/articles/2011/BlanchardEtal_2012_Shifting.pdf.

131. Refugee Act of 1979, Public Law 96-212, *U.S. Statutes at Large* 94 (1980): 102–18.

132. "Mitch Leads to TPS."

133. According to this account, the Mexican government, in turn, adopted "a policy of tolerance" and allowed many to remain in the country. "Aumenta inmigración por Huracán 'Mitch,'" *Reforma* [México D.F., México], February 27, 1999, 12, Gale OneFile: News; "Eleva 'Mitch' repatriaciones," *Reforma* [México D.F., México], March 12, 1999, 11, Gale OneFile: News.

134. Immigration Act of 1990, Public Law 101-649, *U.S. Statutes at Large* 104 (1990): 4978–5088.

135. U.S. Department of Justice, Immigration and Naturalization Service, "The Designation of Honduras under Temporary Protected Status," *Federal Register* 64 (January 5, 1999): 524–26, https://www.govinfo.gov/app/details/FR-1999-01-05/98-34849; U.S.

Department of Justice, Immigration and Naturalization Service, "The Designation of Nicaragua under Temporary Protected Status," *Federal Register* 64 (January 5, 1999): 526–28, https://www.govinfo.gov/app/details/FR-1999-01-05/98-34848.

136. U.S. Department of Justice, Immigration and Naturalization Service, "News Release: INS Grants Twelve-Month Extension of Temporary Protected Status (TPS) for Eligible Hondurans and Nicaraguans," May 5, 2000, Clinton Digital Library. In the official request for extension of TPS, the Honduran government stated that the INS reported that 96,000 Hondurans had requested TPS by 2000 and 83,000 had received work authorization. See "Request for Extension of Designation of Honduras under Temporary Protected Status (TPS) Program," March 1, 2000, 5, Clinton Digital Library.

137. "Guatemala Given Little Chance of Temporary Protected Status for Its Citizens in the U.S.," Latin America Data Base, May 1, 2008, https://digitalrepository.unm.edu/cgi/viewcontent.cgi?article=10603&context=noticen.

138. Email, Theodore J. Piccone to James F. Babbitt and Richard M. Saunders, "FW: TPS backgrounder for POTUS mission to Central America [unclassified]," November 8, 1998, in file "Exchange-Record (Sept 97–Jan 01) [06/11/1998–12/28/1998]," 2011-1044-F, box 6, Clinton Presidential Library; email, Joshua Gotbaum, February 8, 1999, forwarded by Martha Foley to Christina L. Burrell, February 8, 1999, in file "WHO [Detention, INS, Border Patrol] [06/24/1994–07/02/1999]," 2011-1044-F, box 1, Clinton Presidential Library; email, Steven M. Mertens to Leanne A. Shimabukuro, February 9, 1999, in file "Shimabukuro, Leanne, Crime/INS [Immigration and Naturalization Service] Detention," 2011-1044-F, box 1, Clinton Presidential Library.

139. Email, Scott W. Busby to Richard L. Denniston, "Gobush Q&As with Immigration changes [unclassified]," February 16, 1999.

140. Even legal immigrants who had committed minor offenses years before the law was passed and had served their punishment could be placed in deportation proceedings. "Due Process: Recent Immigration Laws Go Too Far," in file "Irene Bueno, Detention/Discretion-Expedited Removal," 2011-1044-F, box 1, Clinton Presidential Library.

141. "Due Process: Recent Immigration Laws Go Too Far"; email, Valerie N. Guarnieri to Philip C. Bobbitt et al., "FW: Gobush Q&As with immigrant changes [unclassified]," February 6, 1999.

142. "Central America: Hurricane Mitch," *Migration News*, University of California, Davis, December 1998, https://migration.ucdavis.edu/mn/more.php?id=3451.

143. "Central America: Hurricane Mitch"; Human Rights Defense Center, "News in Brief," *Prison Legal News*, August 15, 1999, https://www.prisonlegalnews.org/news/1999/aug/15/news-in-brief/.

144. Congress first created the concept of aggravated felony in the Anti-Drug Abuse Act of 1988. "Aggravated Felonies: An Overview," American Immigration Council, December 16, 2016, https://www.americanimmigrationcouncil.org/research/aggravated-felonies-overview. See also "Aggravated Felonies and Deportation," Transactional Records Access Clearinghouse, accessed January 24, 2019, https://trac.syr.edu/immigration/reports/155/.

145. Email, Steven M. Mertens to Leanne A. Shimabukuro, February 9, 1999; email, Joshua Gotbaum, February 8, 1999, forwarded by Martha Foley to Christina L. Burrell, February 8, 1999.

146. Text of oral statement to Congress by Mark Schneider, March 2, 1999.

147. Immigration lawyers were also advised to tell their clients filing for asylum that they would be disqualified if they returned to their home countries to assist their families, no matter how short the visit; the courts would interpret their departure to mean that they no longer faced persecution. Sandoval, "In Wake of Mitch, Massive Illegal Immigration Is Expected"; Viglucci, "Influx of Refugees after Hurricane Mitch Now Seems Unlikely."

148. A year later the numbers of deportees remained small: only 450 criminal and noncriminal aliens were deported compared with over 13,000 Salvadorans, Guatemalans, and Hondurans. "Monthly Statistical September FY 1999 Year End Report," Immigration and Naturalization Service, October 29, 1999, 3, https://www.dhs.gov/sites/default/files/publications/INS_MonthlyStatisticalReport_YearEnd_1999.pdf; Department of Homeland Security, "Table 65: Aliens Removed by Criminal Status and Region and Country of Nationality, Fiscal Years 1995–2001," *Yearbook of Immigration Statistics 2001*, accessed June 1, 2019, https://www.dhs.gov/immigration-statistics/yearbook/2001.

149. "Silver Lining."

150. Reichman, "Honduras: The Perils of Remittance Dependence and Clandestine Migration"; Jana Sládková, *Journeys of Undocumented Honduran Migrants to the United States* (El Paso: LFB Scholarly Publications, 2010), 9.

151. According to Katharine Donato and Shirin Hakimzadeh, Hondurans made up over 12 percent of the foreign-born population of New Orleans in 1970. Their numbers dropped to 6.7 percent in 1980 but grew to 9.2 and 9.7 percent by 1990 and 2000 respectively. "All classes of Hondurans settled in New Orleans," wrote the authors. "Among the first Hondurans were wealthy young women who were educated with the Ursuline nuns and poor immigrants who worked as mechanics and carpenters at the steamship wharves. Later on, Hondurans and other Central Americans made up the bulk of the service sector working in casinos and restaurants in the New Orleans area. By the time Katrina hit, New Orleans was home to an estimated 140,000 to 150,000 Hondurans, making it the largest Honduran community in the nation." Katharine Donato and Shirin Hakimzadeh, "The Changing Face of the Gulf Coast: Immigration to Louisiana, Mississippi, and Alabama," Migration Policy Institute, January 1, 2006, http://www.migrationpolicy.org/article/changing-face-gulf-coast-immigration-louisiana-mississippi-and-alabama. See also "Louisiana: Demographics and Social," Migration Policy Institute, accessed April 26, 2020, https://www.migrationpolicy.org/data/state-profiles/state/demographics/LA; and Reichmann, "Honduras: The Perils of Remittance Dependence and Clandestine Migration."

152. Sládková, *Journeys of Undocumented Honduran Migrants to the United States*, 9.

153. Donato and Hakimzadeh, "Changing Face of the Gulf Coast."

154. Sládková, *Journeys of Undocumented Honduran Migrants to the United States*, 94.

155. "Monthly Statistical September FY 1999 Year End Report," 3.

156. Daniel R. Reichman, *The Broken Village: Coffee, Migration, and Globalization in Honduras* (Ithaca: ILR Press, 2011), 47.

157. Reichman, 50, 167.

158. "Request for Extension of Designation of Honduras under Temporary Protected Status (TPS) Program," March 1, 2000, 8.

159. Torres Calderón, "Quién conoce Honduras?," 13.

160. Telegram (cable) from Ambassador Larry Leon Palmer, "Scenesetter for Codel Shelby's June 28–30 Visit to Honduras," WikiLeaks, June 22, 2004, https://wikileaks.org/plusd/cables/04TEGUCIGALPA1411_a.html; Reichman, "Honduras: The Perils of Remittance Dependence and Clandestine Migration"; Allen Jennings and Matthew Clarke, "The Development Impact of Remittances to Nicaragua," *Development in Practice* 15, no. 5 (August 2005): 685. By 2005, Salvadoran, Guatemalan, Honduran, and Nicaraguan migrants sent "$5.50 for every dollar that flowed in from big foreign capital" (in Guatemala, the proportion was 14:1). See Jose Luis Rocha, "Remittances in Central America: Whose Money Is It Anyway?," *Journal of World-Systems Research* 17, no. 2 (2011): 468.

161. Telegram (cable) from Ambassador Charles A. Ford, "Honduras Scenesetter for DAS WHA Madison, Visit to Honduras from April 26–28," WikiLeaks, April 21, 2006, https://wikileaks.org/plusd/cables/06TEGUCIGALPA743_a.html; "Central America: USAID Assists Migrants Returning to Their Home Countries, but Effectiveness of Reintegration Efforts Remains to Be Determined," General Accounting Office, November 2018, https://www.gao.gov/assets/700/695298.pdf; "IOM Promotes Reintegration of Returnees in Honduras," IOM, October 22, 2019, https://www.iom.int/news/iom-promotes-reintegration-returnees-honduras; "Assisted Returns," IOM, accessed March 17, 2021, https://triangulonorteca.iom.int/assisted-returns.

162. As an alternative, TPS recipients could also try to secure from the USCIS advance permission (called advance parole) to travel abroad so they could then be paroled into the United States. "Temporary Protected Status: An Overview," American Immigration Council, February 2, 2020, https://www.americanimmigrationcouncil.org/research/temporary-protected-status-overview.

163. U.S. Supreme Court, *Sanchez et ux. v. Mayorkas, Secretary of Homeland Security*, June 7, 2021, https://www.supremecourt.gov/opinions/20pdf/20-315_q713.pdf.

164. "H.R.6—117th Congress (2021–2022): American Dream and Promise Act of 2021," Congress.gov, accessed June 28, 2021, https://www.congress.gov/bill/117th-congress/house-bill/6.

165. Leticia Salomón, *Criminalidad y violencia en Honduras: Retos y desafíos para impulsar la reforma*, Centro de Documentación de Honduras, accessed March 17, 2021, http://www.cedoh.org/resources/Inicio/Cartilla-1.pdf; Elizabeth Ferris, "Criminal Violence and Displacement: Notes from Honduras," Brookings-LSE Project in Internal Displacement, November 8, 2013, http://www.brookings.edu/blogs/up-front/posts/2013/11/08-honduras-violence-displacement-ferris; Vanda Felbab-Brown, "Crime as a Mirror of Politics: Urban Gangs in Indonesia," *Brookings Foreign Policy Trip Reports*, February 6, 2013, http://www.brookings.edu/research/reports/2013/02/06-indonesia-gangs-felbabbrown; Hal Brands, "Crime, Violence, and the Crisis in Guatemala: A Case Study in the Erosion of the State," Strategic Studies Institute, U.S. Army War College, May 2010, https://press.armywarcollege.edu/cgi/viewcontent.cgi?article=1600&context=monographs.

166. According to one report, "Indications of corruption in law enforcement, judicial, and military entities plague counter-narcotics efforts. Despite technical support from the U.S. government, the arrest, prosecution, and incarceration of major narco-trafficking offenders remains problematic." Telegram (cable) from Ambassador Larry Leon Palmer, "Caribbean Basin Economic Recovery Act Report to Congress: Honduras," WikiLeaks, August 19, 2003, https://wikileaks.org/plusd/cables/03TEGUCIGAL-PA1964_a.html. See also Brands, "Crime, Violence, and the Crisis in Guatemala."

167. Some of the deportees were members of U.S.-grown gangs like the Mara Salvatruchas (or MS-13) and Barrio 18, who were targeted for expulsion beginning in 1992 when the INS created the "Violent Gang Task Force" to work with local and state police forces. Street gangs that in California had counted a few hundred members now spread across Central America. They were well-armed and well-funded and had a transnational reach. "Crime and Violence: A Staggering Toll on Central American Development," World Bank, April 7, 2011, http://www.worldbank.org/en/news/press-release/2011/04/07/crime-violence-staggering-toll-central-american-development. See also Cynthia Arnson and Eric L. Olson, eds., *Organized Crime in Central America: The Northern Triangle*, Woodrow Wilson Center Reports on the Americas No. 29, November 2011, https://www.wilsoncenter.org/publication/organized-crime-central-america-the-northern-triangle-no-29; and Jennifer M. Chacon, "Whose Community Shield? Examining the Removal of the Criminal Street Gang Member," *University of Chicago Law Forum*, no. 11 (2007), 325, https://chicagounbound.uchicago.edu/cgi/viewcontent.cgi?article=1412&context=uclf.

168. Sondeo de Opinión Pública, *Percepciones sobre la situación hondureña en el año, 2017*, April 2018, https://www.cedoh.org/Biblioteca_CEDOH/archivo.php?id=1003; D'Vera Cohn, Jeffrey S. Passel, and Ana Gonzalez-Barrera, "Rise in U.S. Immigrants from El Salvador, Guatemala, and Honduras Outpaces Growth from Elsewhere," Pew Research Center, December 7, 2017, https://www.pewresearch.org/hispanic/2017/12/07/rise-in-u-s-immigrants-from-el-salvador-guatemala-and-honduras-outpaces-growth-from-elsewhere/; Cynthia Arnson and Eric L. Olson, "Introduction," *Organized Crime in Central America: The Northern Triangle*, 12.

169. Telegram (cable) from Ambassador Larry Leon Palmer, "Media Reactions on TPS," WikiLeaks, November 3, 2004, https://wikileaks.org/plusd/cables/04TEGUCIGALPA2478_a.html.

170. The Central American Integration System also established a common customs procedure for all of Central America to facilitate easier travel across the region. Telegram (cable) from Deputy Chief of Mission Bruce Wharton, "Central American Presidential Summit," WikiLeaks, July 9, 2004, https://wikileaks.org/plusd/cables/04GUATEMALA1698_a.html.

171. See, for example, the letter from Secretary of State Roberto Flores Bermúdez to Madeleine Albright, March 1, 2000, Clinton Digital Library; and "Request for Extension of Designation of Honduras under Temporary Protected Status (TPS) Program," March 1, 2000. Flores Bermúdez discusses how the rains and flooding have disrupted reconstruction efforts.

172. Telegram (cable) from Ambassador Larry Leon Palmer, "TPS for Honduras—Update on Economy and Reconstruction Efforts," WikiLeaks, February 9, 2004, https://wikileaks.org/plusd/cables/04TEGUCIGALPA293_a.html.

173. Telegram (cable) from Ambassador Larry Leon Palmer, "Embassy Tegucigalpa Recommends Extension of TPS, with Planning for the End Game," WikiLeaks, February 14, 2003, https://wikileaks.org/plusd/cables/03TEGUCIGALPA442_a.html.

174. Maria Cristina Garcia, "The New Asylum Seekers," in *Refugee Challenge in Post–Cold War America*.

175. "DHS Releases End of Year Fiscal Year 2016 Statistics," Department of Homeland Security, December 30, 2016, https://www.dhs.gov/news/2016/12/30/dhs-releases-end-year-fiscal-year-2016-statistics; D'Vera Cohn, Jeffrey S. Passel, and Kristen Bialik, "Many Immigrants with Temporary Protected Status Face Uncertain Future in U.S.," Pew Research Center, November 27, 2019, https://www.pewresearch.org/fact-tank/2019/11/27/immigrants-temporary-protected-status-in-us/.

176. Muzaffar Chishti, Sarah Pierce, and Jessica Bolter, "The Obama Record on Deportations: Deporter in Chief or Not?," Migration Policy Institute, January 26, 2017, https://www.migrationpolicy.org/article/obama-record-deportations-deporter-chief-or-not.

177. "Behind the Headlines: Temporary Protected Status," International Rescue Committee, October 4, 2018, https://www.rescue.org/article/behind-headlines-temporary-protected-status; Nick Miroff, Seung Min Kim, and Joshua Partlow, "U.S. Embassy Cables Warned against Expelling 300,000 Immigrants. Trump Officials Did It Anyway," *Washington Post*, May 8, 2018, https://www.washingtonpost.com/world/national-security/us-embassy-cables-warned-against-expelling-300000-immigrants-trump-officials-did-it-anyway/2018/05/08/065e5702-4fe5-11e8-b966-bfb0d2dad62_story.html?noredirect=on&utm_term=.21e0570a801a; *Tantamount to a Death Sentence: Deported TPS Recipients Will Experience Extreme Violence and Poverty in Honduras and El Salvador*, Centro Presente, Alianza Americas, and Lawyers' Committee for Civil Rights and Economic Justice, July 2018, http://lawyersforcivilrights.org/wp-content/uploads/2018/08/Updated-TPS-Delegation-Report-July-2018.pdf.

178. Nielsen announced that Nicaraguans would lose TPS protections on January 5, 2019, and Hondurans on January 5, 2020. Jill H. Wilson, "Temporary Protected Status: Overview and Current Issues," Congressional Research Service, April 1, 2020, 6, https://fas.org/sgp/crs/homesec/RS20844.pdf; USCIS, "Termination of the Designation of Nicaraguans for Temporary Protected Status," *Federal Register*, 82 (December 15, 2017): 59636–42, https://www.federalregister.gov/documents/2017/12/15/2017-27141/termination-of-the-designation-of-nicaragua-for-temporary-protected-status; USCIS, "Termination of the Designation of Hondurans for Temporary Protected Status," *Federal Register* 83 (June 5, 2018): 26074–80, https://www.federalregister.gov/documents/2018/06/05/2018-12161/termination-of-the-designation-of-honduras-for-temporary-protected-status.

179. Three earthquakes struck El Salvador in 2001. The most devastating occurred in January 2001, when an earthquake measuring 7.1 on the Richter scale devastated parts of that country. A second earthquake in February 2001, measuring 6.6, caused additional damage and loss of life. In total, over 1,000 people died and 1 million were left homeless. The country experiences significant seismic activity (over 2,000 tremors per year): earthquakes measuring over 5.0 struck the country in December 2006, March 2007, January 2010, November 2011, October 2014, May 2018, and May 2019. There are also seven active volcanoes in El Salvador. In October 2005, the country's largest volcano, Llamatepec, erupted twice in a two-day period in the Santa Ana

Department, forcing the evacuation of over 4,850 people. National Geophysical Data Center, "Earthquakes in El Salvador since 1950," WorldData.Info, accessed May 6, 2020, https://www.worlddata.info/america/el-salvador/earthquakes.php; Global Intelligence Files, "El Salvador," WikiLeaks, February 13, 2013, https://wikileaks.org/gifiles/docs/53/5391524_el-salvador-.html.

180. There were 411,326 TPS holders in 2019. Wilson, "Temporary Protected Status: Overview and Current Issues," 5. The National Immigration Forum and International Rescue Committee listed other numbers: 195,000 Salvadorans, 57,000 Hondurans, and 2,550 Nicaraguans. "Fact Sheet: Temporary Protected Status," National Immigration Forum, May 5, 2018, https://immigrationforum.org/article/fact-sheet-temporary-protected-status/; "Behind the Headlines: Temporary Protected Status," International Rescue Committee, July 26, 2018, https://www.rescue.org/article/behind-headlines-temporary-protected-status.

181. Larry O'Connor, "Hey Salvadoran Refugees, What Part of 'Temporary' Don't You Understand?," *Washington Times*, January 9, 2018, https://www.washingtontimes.com/news/2018/jan/9/hey-salvadoran-refugees-what-part-temporary-do-you/.

182. Cecelia Menjívar, "Temporary Protected Status in the United States: The Experiences of Honduran and Salvadoran Immigrants," Center for Migration Research, University of Kansas, May 2017, http://ipsr.ku.edu/migration/pdf/TPS_Report.pdf.

183. "Behind the Headlines: Temporary Protected Status."

184. Ramos v. Nielsen, 336 F. Supp. 3d 1075 (N.D. Cal. 2018), https://casetext.com/case/ramos-v-nielsen-3. DHS required TPS recipients to have reregistered by February 13, 2018.

185. See, for example, America's Voice, "Biden Administration Announces 15 Month Temporary Protected Status (TPS) Extension for El Salvador, Honduras, Nicaragua, Haiti, Nepal, and Sudan," September 9, 2021, https://americasvoice.org/press_releases/biden-administration-announces-15-month-temporary-protected-status-tps-extension-for-el-salvador-honduras-nicaragua-haiti-nepal-sudan/.

186. In addition to the American Dream and Promise Act (H.R. 6), introduced in both 2019 and 2021, the Farm Workforce Modernization Act (introduced in 2019 and 2021) created a renewable "certified agricultural worker" status for TPS recipients who had performed a certain amount of agricultural labor over the previous two years, which could lead to eventual legal permanent residence. Wilson, "Temporary Protected Status: Overview and Current Issues," 15.

187. The U.S. government departments and agencies involved include the Environmental Protection Agency, the Departments of State and Energy, USAID, the USDA, and the NOAA. See "U.S. Country Studies Program: Support for Climate Change Studies, issue no. 1" (August 1994), in file "Country Studies to Address Climate Change," Domestic Policy Council, Brian Burke Climate Mitigation Group [2] to Domestic Gas and Oil Initiative [1], 2013-1074-5, box 11, Clinton Presidential Library.

188. Telegram from Chargé d'Affaires ad interim Robert I. Blau, "SERVIR Project Supports Disaster Relief and Environmental Monitoring in Central America," WikiLeaks, May 18, 2009, https://wikileaks.org/plusd/cables/09SANSALVADOR433_a.html; "SERVIR Overview," NASA, accessed May 10, 2020, https://www.nasa.gov/mission_pages/servir/overview.html.

189. Rodgers, "In Debt to Disaster."

190. Rodgers, "In Debt to Disaster."

191. "The World Bank in Honduras," World Bank, accessed April 26, 2020, https://www.worldbank.org/en/country/honduras/overview; Marco Antonio Hernández Ore, Liliana D. Sousa, and J. Humberto López, "Honduras: Unlocking Economic Potential for Greater Opportunities," World Bank, accessed April 26, 2020, http://documents.worldbank.org/curated/en/519801468196163960/pdf/103239-v2-PUB-P151906-Box394858B-PUBLIC-DOI-10-1596K8570-EPI-K8570.pdf.

192. "The World Bank in Nicaragua," World Bank, accessed April 26, 2020, https://www.worldbank.org/en/country/nicaragua/overview; International Development Association, "Nicaragua: Supporting Progress in Latin America's Second Poorest Country," World Bank, August 2009, 1–2, http://siteresources.worldbank.org/IDA/Resources/IDA-Nicaragua.pdf. Hurricane Mitch caused an inflation rate of 18.5 percent. See telegram (cable) from Ambassador Paul A. Trivelli, "Nicaragua: 2007 Inflation Rate May Reach 16.5%," WikiLeaks, December 1, 2007, https://wikileaks.org/plusd/cables/07MANAGUA2524_a.html.

193. On the topic of trafficking, see Instituto Internacional de Derechos Humanos, *Esclavitud moderna: Tráfico sexual en las Américas*, DePaul University, June 2003, https://law.depaul.edu/about/centers-and-institutes/international-human-rights-law-institute/publications/Documents/report_spanish.pdf; Olivia Ruiz Marrujo, "Los riesgos de cruzar: La migración centroamericana en la frontera Mexico-Guatemala," *Frontera norte* 13, no. 25 (January 2001), 7–41; *Honduras: Assessment of the Damage Caused by Hurricane Mitch, 1998*, 69; Rodgers, "In Debt to Disaster"; "Scenesetter for SECDEF Rumfeld's Visit to Honduras August 20," WikiLeaks, August 15, 2003, https://wikileaks.org/plusd/cables/03TEGUCIGALPA1931_a.html; and Karen Lowe, "Hurricane Mitch Puts Tens of Thousands of Children Out in Streets," Agence France-Presse, November 12, 1998, Clinton Digital Library.

194. José Romero Reynó, Mauro Castelló González, and Imti Choonara, "Child Health in Central America and the Caribbean," *Archives of Disease in Childhood* 100, suppl. 1 (2015): S70–S71, https://doi.org/10.1136/archdischild-2014-306855.

195. Over the next two decades, school attendance and literacy rates improved, however: by 2015, 91 percent of Nicaraguans ages fifteen to twenty-four were literate, and by 2018, over 96 percent of Hondurans ages fifteen to twenty-four were literate. In terms of overall numbers for the general population, however, both Honduras and Nicaragua had among the highest illiteracy rates in Central America. "Nicaragua," UNESCO Institute for Statistics, accessed January 21, 2021, http://uis.unesco.org/en/country/ni; "Honduras," UNESCO Institute for Statistics, accessed January 21, 2021, http://uis.unesco.org/en/country/hn.

196. Cited in Fundación Inter-Americana, "Honduras: Reflexiones sobre el huracán Mitch y sus secuelas."

197. Rodgers, "In Debt to Disaster."

198. José Antonio Milán Pérez, *Apuntes sobre el cambio climático en Nicaragua* (Managua: Pascal Chaput, 2010), https://coin.fao.org/coin-static/cms/media/5/12802494073060/apuntes_sobre_cambio_climatico_en_nicaragua.pdf; Comisión Económica para América Latina y el Caribe (CEPAL), *Nicaragua: Efectos*

del cambio climático sobre la agricultura (Mexico: CEPAL, 2010), https://www.cepal.org/es/publicaciones/25925-nicaragua-efectos-cambio-climatico-la-agricultura; Joint Global Change Research Institute, Battelle Memorial Institute, Pacific Northwest Division, and National Intelligence Council, *Mexico, the Caribbean, and Central America*.

199. Cited in Rodgers, "In Debt to Disaster."

200. According to the U.S. embassy, the co-organizers were Roger Escober, Bertha Oliva de Nativi, Jorge Varela, P. Osmin Flores, Walter E. Ulloa, Angel Amilcar Colon, and Rufino Rodriguez.

201. Among those murdered were Jeannette Kawas, Carlos Escaleras, Carlos Antonio Luna López, and Carlos Roberto Flores. Telegram (cable) from U.S. Ambassador Larry Leon Palmer, "March for Life Focused on Forests Energizes Honduran Environmental Movement," WikiLeaks, July 16, 2004, https://wikileaks.org/plusd/cables/04TEGUCIGALPA1581_a.html.

202. Telegram (cable) from U.S. Ambassador Larry Leon Palmer, "March for Life Focused on Forests Energizes Honduran Environmental Movement," July 16, 2004. Also in attendance at the march was former U.S. ambassador to El Salvador Bob White, who headed the U.S. embassy in San Salvador during 1980–81 and now led a delegation from the Center for International Policy, a Washington-based think tank.

203. Beckman, "Labyrinth of Deception." According to Beckman, "[Secretary of State Hillary] Clinton concealed the preplanned and criminal nature of the coup as it unfolded, at the time when an official declaration would have had the most potentially significant impact. She repeatedly dismissed internal warnings that she was letting a dangerous and corrupt regime succeed, during the most critical time when it could have been reversed." See also Dana Frank, "WikiLeaks Honduras: U.S. Linked to Brutal Businessman," *The Nation*, October 21, 2011.

204. Yolanda Polo, "Entrevista con Berta Oliva-COFADEH, Honduras: Honduras se encuentra en estado de S.O.S. internacional en materia de derechos humanos," Coordinadora de Organizaciones Para el Desarrollo, June 6, 2013, https://coordinadoraongd.org/2013/06/entrevista-con-berta-oliva-cofadeh-honduras-honduras-se-encuentra-en-estado-de-s-o-s-internacional-en-materia-de-derechos-humanos/; Beckman, "Labyrinth of Deception."

205. Center for Constitutional Rights and International Federation for Human Rights, "Impunity in Honduras for Crimes against Humanity between 28 June 2009 and 31 October 2012, Submission Pursuant to Article 15 of the Rome Statute of the International Criminal Court," Center for Constitutional Rights, November 12, 2012, https://ccrjustice.org/sites/default/files/assets/files/Honduras%20ICC%20Submission.pdf. See also Nina Lakhani, *Who Killed Berta Cáceres? Dams, Death Squads, and an Indigenous Defender's Battle for the Planet* (London: Verso, 2020); and Dana Frank, *The Long Honduran Night: Resistance, Terror, and the United States in the Aftermath of the Coup* (Chicago: Haymarket Books, 2018). Global Witness reported that at least one of the murderers had received training at the U.S. Army School of the Americas, which had trained some of the assassins of the proxy wars of the 1980s. "Honduras: The Most Dangerous Country in the World for Environmental Activism," Global Witness, January 31, 2017, https://www.globalwitness.org/en/campaigns/environmental-activists/honduras-deadliest-country-world-environmental-activism/.

206. Herman Rosa, "Una iniciativa de restauración a gran escala para construir en el Triángulo Norte, la infraestructura verde y social para un desarrollo sostenible resiliente al clima," Woodrow Wilson International Center for Scholars, the Ford Foundation, and PRISMA, April 16, 2021, https://www.wilsoncenter.org/event/challenges-climate-change-central-america-opportunities-biden-administration; Iliana Monterroso, "Is There a Role for Forests in Addressing the Root Causes of Migration in the Northern Triangle?," Woodrow Wilson International Center for Scholars, the Ford Foundation, and PRISMA, April 16, 2021, https://www.wilsoncenter.org/event/challenges-climate-change-central-america-opportunities-biden-administration.

207. According to Beckman, following the coup, the United States committed over $57 million in direct military aid over the next five years and "rewarded the heavily-compromised military and police with a total of $200 million." Beckman, "Labyrinth of Deception."

208. "Honduras: The Most Dangerous Country in the World for Environmental Activism."

209. REDD+ rewards reforestation projects that absorb carbon emissions but in the past has counted monoculture plantations as a form of reforestation. Danielle Marie Mackey, "Bajo Aguán's Modern Tragedy of the Commons," *Guernica*, December 3, 2012, https://www.guernicamag.com/bajo-aguans-modern-tragedy-of-the-commons/.

210. Brundtland Commission, *Report of the World Commission on Environment and Development: Our Common Future* (Oxford: Oxford University Press, 1987), 16, http://www.un-documents.net/our-common-future.pdf.

211. According to Rodolfo Casillas, this has provoked a backlash not because they are climate refugees but because they are migrants. Rodolfo Casillas, "Migración internacional y cambio climático: Conexiones y desconexiones entre México y Centroamérica," *Urvio* 26 (February 1, 2020): 73–92, doi:10.17141/urvio.26.2020.4038.

212. Hillary Rodham Clinton "Talking It Over," November 11, 1998, in Laura Schiller's Files 1993–2001, Records of the First Lady's Office, 1993–2001, Clinton Presidential Library.

213. Torres Calderón, "Quién conoce Honduras?," 12.

214. Jacobo García, "La zona cero del cambio climático en América Latina," *El País*, February 8, 2020, https://elpais.com/sociedad/2020/02/08/actualidad/1581121631_785715.html.

Chapter 3

1. Dominica's prime minister communicated the impacts of the 160 mph winds on his country of 70,000 people via a series of Facebook posts. See Aubrey Allegretti, "Hurricane Maria: Dominica PM Rescued amid 'Mind-Boggling' Carnage," *SkyNews*, September 19, 2017, https://news.sky.com/story/hurricane-maria-dominica-pm-rescued-amid-mind-boggling-carnage-11042947.

2. The National Hurricane Center called Irma the strongest hurricane ever recorded in the Atlantic basin, outside of the Caribbean Sea and the Gulf of Mexico. The destruction in Barbuda forced a complete evacuation of the population, most of whom settled in nearby Antigua. Hurricane Irma was the first major hurricane to make landfall in

Florida since 2005 and forced one of the largest evacuation missions in Florida history. Irma also affected the states of Alabama, Georgia, North Carolina, South Carolina, and Tennessee. In addition to Dominica, USVI, and Puerto Rico, Hurricane Maria affected the Dominican Republic, Haiti, Turks and Caicos, the Bahamas, and the southeastern and mid-Atlantic region of the United States. "2017 Hurricane Season FEMA After-Action Report," Federal Emergency Management Agency (FEMA), July 12, 2018, v, https://www.fema.gov/sites/default/files/2020-08/fema_hurricane-season-after-action-report_2017.pdf.

3. Brian Resnick, "Why Hurricane Maria Is Such a Nightmare for Puerto Rico," Vox, September 22, 2017, https://www.vox.com/science-and-health/2017/9/21/16345176/hurricane-maria-2017-puerto-rico-san-juan-meteorology-wind-rain-power; Robinson Meyer, "What's Happening with the Relief Effort in Puerto Rico?," The Atlantic, October 4, 2017, https://www.theatlantic.com/science/archive/2017/10/what-happened-in-puerto-rico-a-timeline-of-hurricane-maria/541956/.

4. "Hurricane Maria Top 10 Strongest in Atlantic and Strongest to Hit Puerto Rico," Just in Weather, accessed March 13, 2019, https://justinweather.com/2017/09/19/hurricane-maria-top-10-strongest-in-atlantic-and-strongest-to-hit-puerto-rico/.

5. "National Guard Plans Relief Efforts as Hurricane Maria Slams Puerto Rico," U.S. Department of Defense, September 20, 2017, https://dod.defense.gov/News/Article/Article/1318026/national-guard-plans-relief-efforts-as-hurricane-maria-slams-puerto-rico/.

6. "National Guard Plans Relief Efforts as Hurricane Maria Slams Puerto Rico."

7. The U.S. Northern Command assisted with "search and rescue operations; port and airfield assessment; emergency route clearance; air and ground transportation; aeromedical evacuation . . . strategic airlift to transport personnel, relief supplies, and equipment; imagery; life-sustaining commodity (e.g., food and water) distribution; power restoration and distribution; temporary shelters; water purification; logistics support; maritime freight support; Federal Aviation Administration (FAA)–capable radars; and installation support bases and responder support camps for FEMA responders." U.S. House of Representatives, Subcommittee on the Interior, Energy, and Environment of the Committee on Oversight and Government Reform, *The Historic 2017 Hurricane Season: Impacts on the U.S. Virgin Islands*, 115th Cong., 2nd sess., March 12, 2018 (Washington: U.S. Government Publishing Office, 2018), 52–55 (hereafter *Historic 2017 Hurricane Season*).

8. "President Donald J. Trump Signs Emergency Declaration for Puerto Rico," FEMA, September 18, 2017, https://www.fema.gov/news-release/2017/09/18/president-donald-j-trump-signs-emergency-declaration-puerto-rico; "President Donald J. Trump Approves U.S. Virgin Islands Emergency Declaration," The White House website [archived], September 21, 2017, https://trumpwhitehouse.archives.gov/briefings-statements/president-donald-j-trump-approves-u-s-virgin-islands-disaster-declaration-2/.

9. *Historic 2017 Hurricane Season*, 97, 99; Jeremy W. Peters, "In the Virgin Islands, Hurricane Maria Drowned What Irma Didn't Destroy," *New York Times*, September 27, 2017, https://www.nytimes.com/2017/09/27/us/hurricane-maria-virgin-islands.html.

10. Meyer, "What's Happening with the Relief Effort in Puerto Rico?"

11. Resnick, "Why Hurricane Maria Is Such a Nightmare for Puerto Rico"; Christopher Gillette, "Aid Begins to Flow in Hurricane-Hit Puerto Rico," Associated Press, September 24, 2017, https://apnews.com/06f5077aff384e508e2f2324dae4eb2e.

12. The Robert T. Stafford Disaster Relief and Emergency Assistance Act signed into law on November 23, 1988, amended the Disaster Relief Act of 1974. The law requires the governor of the affected state to request the official declaration. This declaration is also expected of the governors of the District of Columbia, Puerto Rico, the Virgin Islands, Guam, American Samoa, and the Commonwealth of the Northern Mariana Islands. The Republic of Marshall Islands and the Federated States of Micronesia are also eligible to request a declaration and receive assistance through the Compacts of Free Association. As a result of the Sandy Recovery Improvement Act, federally recognized Indian tribal governments also have the option of pursuing a presidential declaration. See "The Disaster Declaration Process," FEMA, accessed June 13, 2019, https://www.fema.gov/disaster-declaration-process. For information on the Stafford Act, see "Robert T. Stafford Disaster Relief and Emergency Assistance Act, as Amended, and Related Authorities as of June 2019," FEMA, accessed February 22, 2020, https://www.fema.gov/media-library/assets/documents/15271.

13. Gillette, "Aid Begins to Flow in Hurricane-Hit Puerto Rico."

14. "DoD Continues Round-the-Clock Support Following Hurricanes in the Caribbean," U.S. Department of Defense, September 25, 2017, https://dod.defense.gov/News/Article/Article/1323530/dod-continues-round-the-clock-support-following-hurricanes-in-caribbean/.

15. "U.S. Officials Outline Hurricane Relief Efforts," U.S. Department of Defense, September 22, 2017, https://dod.defense.gov/News/Article/Article/1321860/military-officials-outline-hurricane-relief-efforts/; *Historic 2017 Hurricane Season*, 58–60.

16. Inspections are carried out by the state or territorial authorities wherever the dams are located. See Steven Mufson, "Failing Puerto Rico Dam That Threatens Thousands Not Inspected since 2013," *Washington Post*, September 26, 2017, https://www.washingtonpost.com/business/economy/failing-puerto-rico-dam-that-endangers-thousands-not-inspected-since-2013/2017/09/26/cfd26272-a225-11e7-b14f-f41773cd5a14_story.html?utm_term=.6ec5e6567fe0.

17. "DoD, Partner Agencies Support Puerto Rico, Virgin Islands Hurricane Relief Efforts," U.S. Department of Defense, September 26, 2017, https://dod.defense.gov/News/Article/Article/1325245/dod-partner-agencies-support-puerto-rico-virgin-islands-hurricane-relief-efforts/.

18. "Caribbean Area Agriculture, Watershed Recovery One Year Post-Maria," Natural Resources Conservation Service, U.S. Department of Agriculture, accessed January 26, 2020, https://www.nrcs.usda.gov/wps/portal/nrcs/detailfull/pr/newsroom/features/?cid=nrcseprd1420889; Jim Wyss, "Lin-Manuel Miranda Is Trying to Revive Puerto Rico's Battered Coffee Industry," *Miami Herald*, October 24, 2018, https://www.miamiherald.com/news/nation-world/world/americas/article220478580.html.

19. Testimony of Rep. Jenniffer González-Colón, November 8, 2017, *Congressional Record* 163, no. 82, https://www.puertoricoreport.com/wp-content/uploads/2017/11/CREC-2017-11-08-house.pdf; Cornelia Hall, Robin Rudowitz, Samantha

Artiga, and Barbara Lyons, "One Year after the Storms: Recovery and Health Care in Puerto Rico and the U.S. Virgin Islands," Kaiser Family Foundation Issue Brief, September 19, 2018, https://www.kff.org/report-section/one-year-after-the-storms-recovery-and-health-care-in-puerto-rico-and-the-u-s-virgin-islands-issue-brief/; Joshua Hoyos, "Only 25 Customers Still without Power in Puerto Rico, Nearly 11 Months after Maria," *ABC News*, August 7, 2018, https://abcnews.go.com/US/25-customers-power-puerto-rico-11-months-maria/story?id=57072951.

20. Conditions prior to 2017 led the Environmental Protection Agency to file suit against the Puerto Rico Aqueduct and Sewer Authority for its violations of the Clean Water Act. "Puerto Rico Aqueduct and Sewer Authority, et al., Clean Water Act Settlement," EPA, accessed August 6, 2019, https://www.epa.gov/enforcement/puerto-rico-aqueduct-and-sewer-authority-et-al-clean-water-act-settlement; "Puerto Rico Aqueduct and Sewer Authority Settlement," EPA, accessed August 6, 2019, https://www.epa.gov/enforcement/puerto-rico-aqueduct-and-sewer-authority-settlement.

21. Carmen Heredia Rodríguez, "Water Quality in Puerto Rico Remains Unclear Months after Hurricane Maria," *PBS News Hour*, June 14, 2018, https://www.pbs.org/newshour/health/water-quality-in-puerto-rico-remains-unclear-months-after-hurricane-maria.

22. "DoD, Partner Agencies Support Puerto Rico, Virgin Islands Hurricane Relief Efforts"; "DoD Accelerates Hurricane Relief, Response Efforts in Puerto Rico," U.S. Department of Defense, September 30, 2017, https://dod.defense.gov/News/Article/Article/1330501/dod-accelerates-hurricane-relief-response-efforts-in-puerto-rico/; testimony of Rep. Jenniffer González-Colón, November 8, 2017.

23. Drawing on past census data, George Washington University researchers first estimated what the death rate in Puerto Rico would have been between September 2017 and February 2018 had the hurricane not happened at all and then accounted for population change caused by migration. They concluded that between 2,658 and 3,290 excess deaths took place between September and February (with 2,975 representing the midpoint). See Milken Institute School of Public Health, George Washington University, *Ascertainment of the Estimated Excess Mortality from Hurricane María in Puerto Rico*, August 29, 2018, https://publichealth.gwu.edu/sites/default/files/downloads/projects/PRstudy/Acertainment%20of%20the%20Estimated%20Excess%20Mortality%20from%20Hurricane%20Maria%20in%20Puerto%20Rico.pdf. Using surveys, Harvard's T. H. Chan School of Public Health estimated there had been 4,645 excess deaths in Puerto Rico in a roughly three-month period after Hurricane Maria. See Nishant Kishore et al., "Mortality in Puerto Rico after Hurricane Maria," *New England Journal of Medicine*, July 12, 2018, https://www.nejm.org/doi/full/10.1056/NEJMsa1803972. See also "Press Release: AG Walker: Five Hurricane-Related Deaths Recorded in the Territory," United States Virgin Islands Department of Justice, October 2, 2017, http://usvidoj.codemeta.com/documents/AG%20Walker%20said%20five%20hurricane-related%20deaths%20recorded%20in%20the%20Territory.pdf; "Poor, Elderly Puerto Ricans Faced Risk of Dying in the Six Months after Hurricane Maria," Gale OneFile, October 18, 2017, https://link.gale.com/apps/doc/A558341103/STND?u=cornell&sid=STND&xid=bfe2bc96; and Carole Ann Moleti, Lucille Contreras Sollazzo, Marc Minick, and Carol Lynn Esposito, "Aftermath of Maria: An Ethnographic Review of Rescue, Recovery, Climate, and Social Justice in Puerto Rico," *Journal of the New York State Nurses Association*, no. 1 (2020): 45–57.

24. Sandra D. Rodríguez Cotto, "WAPA Radio: Voices amid the Silence and Desperation," in *Aftershocks of Disaster: Puerto Rico before and after the Storm*, ed. Yarimar Bonilla and Marisol LeBrón (Chicago: Haymarket Books, 2019), 70.

25. See Peter Shin, Jessica Sharac, Rachel Gunsalus, Brad Leifer, and Sara Rosenbaum, *Puerto Rico's Community Health Centers: Struggling to Recover in the Wake of Hurricane Maria*, Policy Issue Brief #50 (Washington, D.C.: Milken Institute School of Public Health, November 2017), 10, https://publichealth.gwu.edu/sites/default/files/downloads/GGRCHN/GG%3ARCHN%20Policy%20Issue%20Brief%20%2350%20FV.pdf.

26. Hurricane Maria affected five of the top ten drug manufacturers and eleven of the top twenty products in the world. Kristy Malacos, "Drug Shortages: Contributing Factors, Mitigating the Impact," *Pharmacy Times*, March 21, 2019, https://www.pharmacytimes.com/news/drug-shortages-continue-to-be-a-challenge; Julia Carrie Wong, "Hospitals Face Critical Shortage of IV Bags due to Puerto Rico Hurricane," *The Guardian*, January 10, 2018, https://www.theguardian.com/us-news/2018/jan/10/hurricane-maria-puerto-rico-iv-bag-shortage-hospitals; Lauren Weber, "Hurricane Maria's Effects on the Health Care Industry Is Threatening Lives across the U.S.," *HuffPost*, September 24, 2018, https://www.huffpost.com/entry/iv-bag-drug-shortage-puerto-rico-hurricane-maria_n_5ba1ca16e4b046313fc07a8b.

27. Marina Leonard, "Mapp: Virgin Islanders Deserve Same Aid as Other Americans," *St. John Source*, December 5, 2017, https://stjohnsource.com/2017/12/05/mapp-virgin-islanders-deserve-same-aid-as-other-americans/.

28. *Historic 2017 Hurricane Season*, 24–25, 75–84, 85–86; Michael Herzenberg, "U.S. Virgin Islands Governor to Meet with Trump, Discuss Federal Relief Aid," Spectrum News, October 2, 2017, https://www.ny1.com/nyc/all-boroughs/news/2017/10/2/us-virgin-islands-governor-kenneth-mapp-to-meet-president-donald-trump-relief-aid; Brian O'Connor, "Mapp Gives Trump $750-Million First Guess on Cost of Rebuilding," *Virgin Islands Daily News*, October 4, 2017, http://www.virginislandsdailynews.com/breaking/mapp-gives-trump-750-million-first-guess-on-cost-of-rebuilding/article_6e63f4a2-8893-5cb6-a34b-6178f4ce9586.html.

29. *Historic 2017 Hurricane Season*, 20–21, 84–86.

30. Leonard, "Mapp: Virgin Islanders Deserve Same Aid as Other Americans"; *Historic 2017 Hurricane Season*, 31, 74, 85–86.

31. Bernetia Akin, "Study: More USVI Health Damage from Irmaria Than Official Stats," *St. John Source*, November 6, 2019, https://stjohnsource.com/2019/11/06/report-hundreds-may-have-died-due-to-storm-damage-to-v-i-health-system/. See also Muhammad Abdul Baker Chowdhury et al., "Health Impacts of Hurricanes Irma and Maria on St. Thomas and St. John, U.S. Virgin Islands, 2017–2018," *American Journal of Public Health* 109 (December 2019): 1725–32; *Historic 2017 Hurricane Season*, 25.

32. *Historic 2017 Hurricane Season*, 90. Other sources placed the number at closer to a thousand, including caregivers. The National Disaster Medical System, a partnership between the Departments of Health and Human Services, Homeland Security, Defense, and Veterans Affairs, helped cover the cost of care and housing for 803 of the evacuees. Most of the evacuees were unable to return home and have either died or moved in with friends or relatives in the fifty states. Megan Jula, "Hurricanes Drove

More Than 1,000 Medical Evacuees from the Virgin Islands. Many Can't Go Home," *Mother Jones*, June 1, 2018, https://www.motherjones.com/environment/2018/06/hurricanes-drove-more-than-1000-medical-evacuees-from-the-virgin-islands-many-cant-go-home/; Peters, "In the Virgin Islands, Hurricane Maria Drowned What Irma Didn't Destroy."

33. Centers for Disease Control and Prevention, "Shouting in the Dark: Emergency Communication in USVI after Irma and Maria," *Public Health Matters Blog*, March 12, 2018, https://blogs.cdc.gov/publichealthmatters/2018/03/usvi/.

34. Akin, "Study: More USVI Health Damage from Irmaria Than Official Stats"; Greg Allen, "After 2 Hurricanes, a Floodgate of Mental Health Issues in the U.S. Virgin Islands," National Public Radio, April 23, 2019, https://www.npr.org/2019/04/23/716089187/after-two-hurricanes-a-floodgate-of-mental-health-issues-in-the-virgin-islands.

35. Hall et al., "One Year After the Storms."

36. According to Rodríguez Cotto, Puerto Rico has more radio stations per square mile than any other U.S. territory and has the second oldest station in Latin America. Rodríguez Cotto, "WAPA Radio," 66.

37. "Find Updated Information about Your Town in This Puerto Rico Search Engine," Univision News, accessed June 15, 2019, https://www.univision.com/univision-news/united-states/find-updated-information-about-your-town-in-this-puerto-rico-search-engine?hootPostID=3a441a1f7385df653448d79137548bf0#; "#PRActívate," Google Doc, accessed June 15, 2019, https://docs.google.com/forms/d/e/1FAIpQLSfUfYfrS7rw4omTrMqE6yQBbZ9FGFB7uMhm2zpQmzUSLB1YAA/viewform.

38. See, for example, Morning Consult, National Tracking Poll #170916, September 22–24, 2017, https://morningconsult.com/wp-content/uploads/2017/10/170916_crosstabs_pr_v1_KD.pdf. A USA Today/Suffolk University Political Research Center poll conducted in March 2017 revealed that only 47 percent of those polled believed that Puerto Ricans were citizens by birth. USA Today Poll: March 2017, Question 33, USSUFF.030717.R31, Suffolk University Political Research Center (Cornell University, Ithaca, NY: Roper Center for Public Opinion Research, 2017). See also Alan Gomez, "Yes, Puerto Rico Is Part of the United States," *USA Today*, September 26, 2017, https://www.usatoday.com/story/news/world/2017/09/26/yes-puerto-rico-part-united-states/703273001/.

39. Anushka Sha, Allan Ko, and Fernando Peinado, "The Mainstream Media Did Not Care about Puerto Rico until It Became a Trump Story," *Washington Post*, November 27, 2017, https://www.washingtonpost.com/news/posteverything/wp/2017/11/27/the-mainstream-media-didnt-care-about-puerto-rico-until-it-became-a-trump-story/.

40. Dhrumil Mehta, "The Media Really Has Neglected Puerto Rico," FiveThirtyEight, September 28, 2017, https://fivethirtyeight.com/features/the-media-really-has-neglected-puerto-rico/.

41. National Association of Hispanic Journalists, "We're calling on national/local media to cover devastation in PR w/ equal attention & responsibility that would be given anywhere in U.S.," Twitter, September 25, 2017, https://twitter.com/NAHJ/status/912332139894624258?ref_src=twsrc%5Etfw%7Ctwcamp%5Etweetem

NOTES TO PAGES 103–4

bed%7Ctwterm%5E912332139894624258&ref_url=https%3A%2F%2Ffivethirtyeight.com%2Ffeatures%2Fthe-media-really-has-neglected-puerto-rico%2F. See also Mehta, "Media Really Has Neglected Puerto Rico."

42. Citizen activists in Puerto Rico, for example, formed a clearinghouse called "SOS Puerto Rico" (through the online news platform *Daily Kos*) to disseminate information about the island. "SOS Puerto Rico," *Daily Kos*, accessed June 16, 2019, https://www.dailykos.com/user/SOS%20Puerto%20Rico. See also Pete Vernon, "As the Media Neglected Puerto Rico, Some Journalists Made It Their Mission," *Columbia Journalism Review*, September 25, 2018, https://www.cjr.org/united_states_project/puerto-rico-begnaud-santiago.php.

43. "Possibly Millions of Water Bottles Meant for Hurricane Maria Victims Left on Tarmac in Puerto Rico," *CBS This Morning*, September 13, 2018, https://www.cbsnews.com/news/puerto-rico-water-bottles-possibly-millions-for-hurricane-maria-victims-sitting-on-tarmac/; "Water Meant for Puerto Rican Hurricane Victims Dumped on Farmland," Yahoo News, July 29, 2019, https://news.yahoo.com/water-meant-puerto-rican-hurricane-victims-dumped-farmland-222421463.html; Connor Perrett, "Puerto Ricans Discovered a Warehouse Full of Unused Food, Water, and Supplies from Hurricane Maria, Resulting in the Firing of the Island's Emergency Manager," *Insider*, January 19, 2020, https://www.insider.com/puerto-rico-residents-find-warehouse-full-of-supplies-from-maria-2020-1.

44. According to Joanisabel González, Puerto Ricans filed 1,250 lawsuits by November 2018. Joanisabel González, "Ataponadas las cortes con demandas por el Huracán María," *El nuevo día*, November 6, 2018.

45. Adrian Florida, "In Reversal, FEMA Says It Won't End Puerto Rico Food and Water Distribution Wednesday," National Public Radio, January 31, 2018, https://www.npr.org/sections/thetwo-way/2018/01/31/582050242/lawmakers-urge-fema-to-reconsider-ending-food-aid-for-puerto-rico.

46. Naomi Klein, *The Shock Doctrine: The Rise of Disaster Capitalism* (New York: Metropolitan Books/Henry Holt, 2007).

47. Ana Banana, "St David of Begnaud, Patron St of Puertorricans. May his cellphone battery be forever charged. @DavidBegnaud," Twitter, September 28, 2017, https://twitter.com/missyfu/status/913518449510486018.

48. CBS News Press Release, "CBS News Correspondent David Begnaud Wins George Polk Award for Public Service for Extensive Coverage of Hurricane Maria," Viacom CBS Press Express, February 20, 2018, https://www.viacomcbspressexpress.com/cbs-news-and-stations/releases/view?id=49582; National Puerto Rican Day Parade, "We recognized David Begnaud, CBS News Correspondent as Campeón Puertorriqueño (Puerto Rican Champion) for his extraordinary, award-winning reporting of Hurricane Maria and for documenting the recovery efforts and progress made over the past 8 months . . . ," Facebook, May 26, 2018, https://www.facebook.com/NationalPuertoRicanParade/photos/we-recognize-david-begnaud-cbs-news-correspondent-as-campe%C3%B3n-puertorrique%C3%B1o-puer/1855269494491611/.

49. See, for example, Lili Siri Spira, "The United States . . . and Territories: America's Paradise Lost," *Berkeley Political Review*, November 11, 2017, https://bpr.berkeley.edu/2017/11/17/the-united-states-and-territories-americas-paradise-lost/; Denise

Oliver Vélez, "Are Trump and the Media Ignoring the Virgin Islands Because Most of the Residents Are Black?," *Daily Kos*, October 11, 2017, https://www.dailykos.com/stories/2017/10/11/1705708/-Are-Trump-the-media-ignoring-the-U-S-Virgin-Islands-because-most-of-the-residents-are-black.

50. Jessica Huseman, "The Breakthrough: How Journalists in the Virgin Islands Covered the Disaster Happening to Them," *ProPublica*, November 10, 2017, https://www.propublica.org/podcast/the-breakthrough-how-journalists-in-the-virgin-islands-covered-the-disaster-happening-to-them.

51. *Historic 2017 Hurricane Season*, 84.

52. See, for example, Andrea González-Ramírez, "'We Survived': A Year after Hurricane Maria, 25 Puerto Rican Women Tell Their Story," Refinery29, September 20, 2018, https://www.refinery29.com/en-us/2018/09/209761/hurricane-maria-anniversary-puerto-rico-women-experience; "Puerto Rico, One Year after María," Voice of Witness, accessed June 18, 2019, http://voiceofwitness.org/stories-puerto-rico/; Melinda González, "Mitigating Disaster in Digital Space: DiaspoRicans Organizing after Hurricane Maria," *International Journal of Mass Emergencies and Disasters* 38, no. 1 (March 2020), 43–53.

53. The Puerto Rican College of Physicians and Surgeons organized the health brigades. Patricia Noboa Ortega, "Psychoanalysis as a Political Act after María," in Bonilla and LeBrón, *Aftershocks of Disaster*, 271–328.

54. Following Hurricane Maria, Casa Pueblo became a "solar energy hub" preparing the community for future disruptions. By 2020, the organization had outfitted "an additional 55 homes with solar energy, along with 89 full-size refrigerators in seven solar-powered homes, an agricultural center, an elder home, the fire station, five mini-markets, a restaurant, a pizzeria, a barbershop, two hardware stores, and two rooms of Bosque Escuela (Forest School), which now serves as a popular meeting place for education." Diane Hembree, "Lessons from Puerto Rico: A Grantmaker Offers a Roadmap for a 'Just Recovery,'" *Inside Philanthropy*, accessed March 16, 2020, https://www.insidephilanthropy.com/home/2020/2/25/lessons-from-puerto-rico-a-grantmaker-offers-a-roadmap-for-a-just-recovery. See also Luis Alberto Ferré Rangel, "Casa Pueblo y su gesta de país," *El nuevo día*, January 30, 2021.

55. González-Ramírez, "'We Survived': A Year after Hurricane Maria."

56. José Andrés and Richard Wolffe, *We Fed an Island: The True Story of Rebuilding Puerto Rico, One Meal at a Time* (New York: HarperCollins, 2018); "Plow to Plate Partnerships Program," World Central Kitchen, accessed June 17, 2019, https://www.worldcentralkitchen.org/plowtoplate/; World Central Kitchen website, accessed June 17, 2019, https://www.worldcentralkitchen.org/; Ana Sofía Peláez, "After Hurricane María, Chef José Andrés 'Fed' Puerto Rico; Now He's Focused on Its Food Production," *NBC News*, February 28, 2019, https://www.nbcnews.com/news/latino/after-hurricane-mar-chef-jose-andr-s-fed-puerto-rico-n977271.

57. Celebrities included Jennifer Lopez, Gina Rodriguez, Ricky Martin, Luis Fonsi, Marc Anthony, Jesse & Joy, Pitbull, Fat Joe, Princess Nokia, and Daddy Yankee. Collectively, American celebrities raised or donated at least $135 million in 2017 for relief and recovery efforts for victims of hurricanes Harvey, Irma, and Maria. Adrienne Gibbs, "Celebs Helped Raise at Least $135 Million for Hurricane Relief in 2017,"

Forbes, December 31, 2017, https://www.forbes.com/sites/adriennegibbs/2017/12/31/celebs-helped-raise-at-least-135-million-for-hurricane-relief-in-2017/#2c99ebe34049.

58. "Kenny Chesney's Love for Love City," kennychesney.com, accessed August 17, 2019, https://www.kennychesney.com/loveforlovecity.

59. Excerpt from promotional video, 21USVIDuncanRelief.org., accessed August 17, 2019, http://21usvihurricanehelp.com/. Duncan had firsthand knowledge of how a hurricane could dramatically impact one's life: in 1989, when he was thirteen, Hurricane Hugo damaged his home, school, and even the island's one Olympic-size pool, where he was training to be a competitive swimmer. After Hugo, Duncan abandoned his plans to be a competitive swimmer and turned his attention to basketball.

60. Michael C. Wright, "Tim Duncan Brings Help and Hope to Hometown Hit Hard by Hurricanes," ESPN, October 25, 2017, https://www.espn.com/nba/story/_/id/21137486/tim-duncan-brings-hope-hometown-hit-hard-hurricanes.

61. Meyer, "What's Happening with the Relief Effort in Puerto Rico?"

62. During the fall of 2017, the president mentioned the Virgin Islands only three times through social media (once after Hurricane Irma and twice after Maria). See Trump Twitter Archive, accessed July 1, 2019, http://www.trumptwitterarchive.com/; "President Trump 45Archived," Twitter, accessed August 19, 2021, https://twitter.com/potus45?lang=en; and "Archived Social Media," Donald J. Trump Presidential Library, accessed August 19, 2021, https://www.trumplibrary.gov/research/archived-social-media. When Twitter suspended Trump's Twitter account, the National Archives could no longer access the former president's tweets. Quint Forgey, "National Archives Can't Resurrect Trump's Tweets, Twitter Says," Politico, April 7, 2021, https://www.politico.com/news/2021/04/07/twitter-national-archives-realdonaldtrump-479743.

63. Within nine days, FEMA distributed 5.1 million meals, 4.7 million liters of water, and over 26,000 cots and blankets in Texas and Louisiana. FEMA distributed 6.6 million meals and 4.7 million liters of water in Florida and other southeastern states in the first four days after the storm. See the following on the FEMA website: "Hurricane Harvey Snapshot," September 3, 2017, https://www.fema.gov/news-release/2017/09/03/hurricane-harvey-snapshot; "Federal Government Continues Federal Response to Hurricane Harvey," September 1, 2017, https://www.fema.gov.news-release/2017/09/01/federal-government-continues-response-hurricane-harvey; "Historical Disaster Response to Hurricane Harvey in Texas," September 22, 2017, https://www.fema.gov/news-release/2017/09/22/historic-disaster-response-hurricane-harvey-texas; "Federal Family Continues Response and Relief Operations Following Hurricane Irma," September 14, 2017, https://www.fema.gov/news-release/2017/09/14/federal-family-continues-response-and-relief-operations-following-hurricane.

64. *Historic 2017 Hurricane Season*, 87.

65. Danny Vinik, "How Trump Favored Texas over Puerto Rico," Politico, March 27, 2018, https://www.politico.com/story/2018/03/27/donald-trump-fema-hurricane-maria-response-480557.

66. "Billion Dollar Weather and Climate Disasters: Overview," National Oceanic and Atmospheric Association, accessed May 8, 2019, https://www.ncdc.noaa.gov/billions/overview.

67. Statement of William B. Long, FEMA Administrator, before the House Appropriations Subcommittee on Homeland Security, November 30, 2017, https://www.fema.gov/media-library-data/1512059678335-560f0e64b55da31154e6b3b04c18628d/HAC.PDF; William L. Painter, "2017 Supplemental Disaster Appropriations: Overview," Congressional Research Service, March 20, 2018, https://fas.org/sgp/crs/homesec/R45084.pdf.

68. Karl Herchenroeder, "FEMA 'Tapped Out' from Too Many 'Unprecedented' Disasters," PJ Media, December 11, 2017, https://pjmedia.com/news-and-politics/fema-tapped-many-unprecedented-disasters/. See also Yxta Maya Murray, "FEMA Has Been a Nightmare: Epistemic Injustice in Puerto Rico," *Willamette Law Review* 55, no. 2 (Spring 2019): 321–94. Murray notes how Long and various Republican legislators blamed the slow response on logistics: "the presumably incalculable and epistemically unavailable (a.k.a., unforeseeable) array of topographic and other complexities that created obstacles between the federal government and the people of Puerto Rico."

69. "2017 Hurricane Season FEMA After-Action Report." See also Frances Robles's account of an earlier draft of the FEMA report secured by the *New York Times*: "FEMA Was Sorely Unprepared for Puerto Rico Hurricane, Report Says," *New York Times*, July 12, 2018, https://www.nytimes.com/2018/07/12/us/fema-puerto-rico-maria.html.

70. By April 30, 2018, FEMA had spent $21.2 billion addressing the impacts of these disasters and assisting 4.8 million households—more households than in the previous ten years combined. "2017 Hurricane Season FEMA After-Action Report," v–vi.

71. Patrick Gillespie, Rafael Romo, and Maria Santana, "Puerto Rico Aid Is Trapped in Thousands of Shipping Containers," CNN, September 28, 2017, https://www.cnn.com/2017/09/27/us/puerto-rico-aid-problem/index.html.

72. Christopher Flavelle, "General Who Led Katrina Response Criticizes Puerto Rico Effort," *Stars and Stripes*, September 28, 2017, https://www.stripes.com/news/us/general-who-led-katrina-response-criticizes-puerto-rico-efforts-1.489974; "Retired Lt. Gen. Russel Honore, Who Led Katrina Relief, Slams Response to Puerto Rico," *CBS News*, September 29, 2017, https://www.cbsnews.com/news/retired-lt-gen-russel-honore-who-led-katrina-relief-slams-response-to-puerto-rico/; "Trump Attacks San Juan Mayor over Hurricane Response," *CNN Wire*, September 30, 2017, https://q13fox.com/2017/09/30/trump-attacks-san-juan-mayor-over-hurricane-response/; Vinik, "How Trump Favored Texas over Puerto Rico"; Alex Ward, "The General Who Turned Around the Response in New Orleans after Katrina Thinks We're Failing in Puerto Rico," Vox, September 28, 2017, https://www.vox.com/world/2017/9/28/16379286/puerto-rico-response-trump-honore-interview.

73. Laura M. Quintero, "Confirma mal manejo de FEMA en el Huracán María," *El nuevo día*, October 2, 2020; Patricia Mazzei and Agustin Armendariz, "FEMA Contract Called for 30 Million Meals for Puerto Ricans. 50,000 Were Delivered," *New York Times*, February 6, 2018, https://www.nytimes.com/2018/02/06/us/fema-contract-puerto-rico.html; Tami Abdollah, "AP Exclusive: Big Contracts, No Storm Tarps for Puerto Rico," Associated Press, November 28, 2017, https://www.apnews.com/cbeff1a939324610b7a02b88f30eafbb; Emma Schwartz, "FEMA Tarp Contractor under Investigation for Fraud," *PBS Frontline*, June 14, 2019, https://www.pbs.org/wgbh/frontline/article/fema-tarp-contractor-under-investigation-for-fraud/.

74. Laura Sullivan, "FEMA Blamed Delays in Puerto Rico on Maria; Agency Records Tell Another Story," National Public Radio, June 14, 2018, https://www.npr.org/2018/06/14/608588161/fema-blamed-delays-in-puerto-rico-on-maria-agency-records-tell-another-story.

75. Cited in Murray, "FEMA Has Been a Nightmare: Epistemic Injustice in Puerto Rico," 354–59.

76. Donald J. Trump, Twitter, September 25, 2017, https://twitter.com/realDonaldTrump/status/912478274508423168.

77. Eric Levenson, "3 Storms, 3 Responses: Comparing Harvey, Irma, and Maria," CNN, September 27, 2017, https://www.cnn.com/2017/09/26/us/response-harvey-irma-maria/index.html; Chris Cillizza, "Trump Sent 18 Tweets on Puerto Rico on Saturday. And Made Things a Whole Lot Worse," *CNN Politics*, October 1, 2017, https://www.cnn.com/2017/10/01/politics/trump-tweets-puerto-rico/index.html.

78. NBC News, "'I Am Mad as Hell': San Juan Mayor Carmen Yulin Cruz Criticizes Maria Response," YouTube, September 29, 2017, https://www.youtube.com/watch?v=41hl5RwfOVc; Daniella Diaz, "San Juan Mayor: 'Dammit, This Is Not a Good News Story,'" CNN, September 29, 2017, https://www.cnn.com/2017/09/29/politics/puerto-rico-hurricane-maria-san-juan-mayor-trump-response/index.html.

79. Brandon Carter, "San Juan Mayor Wears 'Help Us, We Are Dying,' Shirt for TV Interview," *The Hill*, September 29, 2017, https://thehill.com/blogs/blog-briefing-room/news/353204-san-juan-mayor-wears-help-us-we-are-dying-shirt-during-cnn.

80. "Fact Check: Trump's Claim about Puerto Rico Hurricane Death Toll," National Public Radio, September 13, 2018, https://www.npr.org/2018/09/13/647559553/fact-check-trumps-claim-about-puerto-rico-hurricane-death-toll; Jacqueline Thomsen, "Trump Hands Out Supplies during Puerto Rico Visit," *The Hill*, October 3, 2017, https://thehill.com/homenews/administration/353689-trump-tells-puerto-ricans-they-dont-need-flashlights-while-most-of; "San Juan Mayor Continues Keeping It Real, Says Trump Tossing Paper Towels 'Terrible and Abominable,'" Remezcla, October 4, 2017, https://remezcla.com/culture/carmen-yulin-cruz-calls-out-trump-visit/; Tamara Keith, "Fact Check: Puerto Rico Was an Incredible, Unsung Success?," *USA Today*, September 12, 2018, https://www.npr.org/2018/09/12/646997771/fact-check-puerto-rico-was-an-incredible-unsung-success.

81. "The Latest: Pences Hand Out Sandwiches in Puerto Rico," *Washington Post*, October 6, 2017, https://www.washingtonpost.com/national/the-latest-virgin-islands-governor-praises-maria-response/2017/10/06/da6a085a-aab0-11e7-9a98-07140d2eed02_story.html; "Vice President Pence Visits Puerto Rico," *Puerto Rico Report*, October 9, 2017, https://www.puertoricoreport.com/vice-president-pence-visits-puerto-rico/#.YF4vvhopCjJ; Brianna Sacks, "People on the U.S. Virgin Islands Say It's a 'Disgrace' Trump Isn't Visiting," BuzzFeed, October 3, 2017, https://www.buzzfeednews.com/article/briannasacks/us-virgin-islands-trump-visit.

82. Kevin Derby, "Florida Delegation Continues to Urge More Help for Puerto Rico, Virgin Islands," *Sunshine State News*, September 27, 2017, http://sunshinestatenews.com/story/florida-delegation-continues-urge-more-help-puerto-rico-virgin-islands.

83. "DHS Signs Jones Act Waiver," Department of Homeland Security, September 8, 2017, https://www.dhs.gov/news/2017/09/08/dhs-signs-jones-act-waiver.

84. Ruth Brown, "Trump Says Shipping Industry Opposed to Waiver for Puerto Rico," *New York Post*, September 27, 2017, https://nypost.com/2017/09/27/trump-says-shipping-industry-opposed-to-waiver-for-puerto-rico/; Natalie Andrews and Paul Page, "Trump Weighs Waiving Law Barring Foreign Ships from Delivering Aid to Puerto Rico," *Wall Street Journal*, September 27, 2017, https://www.wsj.com/articles/lawmakers-seek-waiver-of-law-barring-foreign-ships-from-delivering-aid-to-puerto-rico-1506529999.

85. Letter from Nydia Velázquez et al. to Elaine Duke, Congresswoman Nydia Velázquez website, September 25, 2017, https://velazquez.house.gov/sites/velazquez.house.gov/files/09252017%20FINAL%20Letter%20to%20DHS%20JonesAct%20Federal%20requirements%20waivers.pdf.

86. Leonard, "Mapp: Virgin Islanders Deserve Same Aid as Other Americans."

87. Mitchell Miller, "U.S. Virgin Islands: What about Us?" *WTOPNews*, September 14, 2018, https://wtop.com/government/2018/09/us-virgin-islands-what-about-us/.

88. The Subcommittee on the Interior, Energy and Environment is a subcommittee of the U.S. House Oversight and Government Reform Committee. "While the people of the Virgin Islands are grateful to the various federal agencies for their respective responses and the progress made thus far, I think it is important to afford the residents of our community the opportunity to hold these entities accountable," said Plaskett, explaining the reasons for holding the hearings in the USVI. "And also to hear directly from those responsible, what, if anything, can be done to improve the recovery and ensure the Virgin Islands is rebuilt stronger, smarter and more efficient." "Plaskett to Host Congressional Oversight Hearing on Hurricane Recovery Efforts in the Virgin Islands," Congresswoman Stacey E. Plaskett website, November 29, 2017, https://plaskett.house.gov/news/documentsingle.aspx?DocumentID=262; *Historic 2017 Hurricane Season*, 1–3, 80.

89. Peters, "In the Virgin Islands, Hurricane Maria Drowned What Irma Didn't Destroy"; "Written Testimony of Governor Kenneth E. Mapp of the United States Virgin Islands before the Senate Committee on Energy and Natural Resources," United States Senate Committee on Energy and Natural Resources, November 14, 2017, https://www.energy.senate.gov/public/index.cfm/files/serve?File_id=A2538A49-2953-4BA1-8C94-0807E62050A5; Deirdre Walsh and Kevin Liptak, "Federal Response to Hurricane Maria Slowly Takes Shape," CNN, September 25, 2017, https://www.cnn.com/2017/09/25/politics/puerto-rico-hurricane-maria-aid-donald-trump/index.html.

90. Jim Garamone, "Federal Emergency Response Working Smoothly, Puerto Rico's Governor Says," *U.S. Department of Defense News*, September 30, 2017, https://dod.defense.gov/News/Article/Article/1330489/federal-emergency-response-working-smoothly-puerto-ricos-governor-says/.

91. Doug Mack, "Empire State of Mind," *Slate*, March 15, 2018, https://slate.com/news-and-politics/2018/03/its-time-for-the-u-s-to-have-a-conversation-about-its-overseas-territories.html.

92. Yarimar Bonilla and Marisol LeBrón, eds., *Aftershocks of Disaster: Puerto Rico before and after the Storm* (Chicago: Haymarket Books, 2019), 11.

93. In 1823, then secretary of state John Quincy Adams called Puerto Rico and Cuba "natural appendages to the North American continent." John Quincy Adams, *Writings of John Quincy Adams: 1820–1823* (Westport, CT: Greenwood Press, 1968), 372.

94. The Charter of Autonomy had granted Puerto Ricans the right to their own legislature, constitution, tariffs, monetary system, treasury, and judiciary.

95. Nelson A. Denis, *War against All Puerto Ricans: Revolution and Terror in America's Colony* (New York: Nation Books, 2015), 55.

96. Sam Erman offers an excellent history of the insular cases in *Almost Citizens: Puerto Rico, the U.S. Constitution, and Empire* (Cambridge: Cambridge University Press, 2019). The following two collections offer a wide range of essays on the insular cases and their implications and legacies: Christina Duffy Burnett and Burke Marshall, eds., *Foreign in a Domestic Sense: Puerto Rico, American Expansion, and the Constitution* (Durham: Duke University Press, 2001); and Gerald L. Neuman and Tomiko Brown-Nagin, eds., *Reconsidering the Insular Cases: The Past and Future of the American Empire* (Cambridge, Mass.: Human Rights Program, Harvard Law School, 2015). The Territorial Clause gives Congress "power to dispose of and make all needful rules and regulations respecting the territory or other property belonging to the United States." According to Gerald L. Neuman, territorial powers have been attributed to the Territorial Clause of Article IV of the Constitution, to the war power, to the treaty power, to the power of Congress to admit new states, and to the implied powers of national sovereignty. Gerald L. Neuman, "Constitutionalism and Individual Rights in the Territories," in Burnett and Marshall, *Foreign in a Domestic Sense*, 184.

97. Downes v. Bidwell, 182 U.S. 244 (1901).

98. Efrén Rivera Ramos, "Deconstructing Colonialism: The 'Unincorporated Territory' as a Category of Domination," in Burnett and Marshall, *Foreign in a Domestic Sense*, 106–7.

99. Luis F. Estrella Martínez, *Acceso a la justicia: Derecho humano fundamental* (San Juan, PR: Biblio Services y Ediciones Situm, 2017); Luis F. Estrella Martínez, "Puerto Rico: La evolución de un apartheid territorial," *Revista jurídica de la Universidad Interamericana de Puerto Rico Facultad de Derecho* 52 (August/May 2017–18): 425–30; Juan R. Torruella, *The Supreme Court and Puerto Rico: The Doctrine of Separate and Unequal* (Rio Piedras, PR: University of Puerto Rico, 1985).

100. Jorge Duany, *Puerto Rico: What Everyone Needs to Know* (New York: Oxford University Press, 2017), 49.

101. Rivera Ramos, "Deconstructing Colonialism," 108.

102. Thomas Aitken, *Poet in the Fortress: The Story of Luis Muñoz Marín* (New York: Signet Books, 1964), 60–62.

103. Bailey W. Diffie and Justine Whitfield Diffie, *Porto Rico: A Broken Pledge* (New York: Vanguard Press, 1931), 199–200.

104. The 1900 Foraker Act created the position of resident commissioner, but the American governor appointed the official for a two-year term rather than its current four-year term.

105. Denis, *War against All Puerto Ricans*, 30.

106. Kathryn Krase, "Sterilization Abuse: The Policies behind the Practice," *National Women's Health Network* 21, no. 1 (January/February 1996), 1+ *Gale Academic OneFile*, link.gale.com/apps/doc/A18306594/AONE?u=nysl_sc_cornl&sid=ebsco&xid=5e8e01e1. In the town of Barceloneta alone, 20,000 women were sterilized. See also Denis, *War against All Puerto Ricans*, 33–36. Denis also discusses the use of radiation experimentation by the federal government, including the use of radiation

against nationalist leader Pedro Albizu Campos when he was imprisoned in a federal penitentiary.

107. In *War against All Puerto Ricans*, Denis chronicles the surveillance strategies of the Puerto Rican police and the FBI over the decades. Secret police dossiers called *carpetas* contained information on over 100,000 people.

108. Puerto Rico Foreign Relations Act of 1950, Public Law 81-600, https://govtrackus.s3.amazonaws.com/legislink/pdf/stat/64/STATUTE-64-Pg319.pdf. The preamble reads, "Be it enacted by the Senate and House of Representatives of the United States of America in Congress assembled, That, fully recognizing the principle of government by consent, this Act is now adopted in the nature of a compact so that the people of Puerto Rico may organize a government pursuant to a constitution of their own adoption." Public Law 82-447 officially approved the constitution.

109. Powers that administer non-self-governing territories are required to transmit information to the UN Decolonization Committee under the terms of Article 73(e), but the new commonwealth status meant the United States was freed from these responsibilities. See Christina Duffy Burnett and Burke Marshall, "Between the Foreign and the Domestic: The Doctrine of Territorial Incorporation, Invented and Reinvented," in Burnett and Marshall, *Foreign in a Domestic Sense*, 19–20.

110. Duany, *Puerto Rico*, 74; Burnett and Marshall, "Between the Foreign and the Domestic," 18; "The Origin of the 'Commonwealth' Label," *Puerto Rico Report*, April 20, 2011, https://www.puertoricoreport.com/the-origin-of-the-commonwealth-label/#.XTMwxHt7lZ2.

111. José Trías Monge, "Injustice according to the Law: The *Insular Cases* and Other Oddities," in Burnett and Marshall, *Foreign in a Domestic Sense*, 233–37. See also José Trías Monge, *Puerto Rico: The Trials of the Oldest Colony in the World* (New Haven: Yale University Press, 1997).

112. In the introductory essay to their anthology, editors Christina Duffy Burnett and Burke Marshall argue that the enhanced commonwealth status has the greater support. They note that "no one today defends the colonial status sanctioned by these [insular] cases, yet the idea of a relationship to the United States that is somewhere 'in between' that of statehood and independence—somehow both 'foreign' and 'domestic' (or neither) has not only survived but enjoys substantial support. A territorial status born of colonialism has been appropriated by colonial subjects." See Burnett and Marshall, "Between the Foreign and the Domestic," in Burnett and Marshall, *Foreign in a Domestic Sense*, 2.

113. D. Andrew Austin, "Economic and Fiscal Conditions in the U.S. Virgin Islands," Congressional Research Service, June 20, 2018, 3, https://www.hsdl.org/?view&did=812532.

114. Rodríguez Cotto, "WAPA Radio," 70.

115. The Puerto Rican legislature passed the Export Services Act and the Individual Investors Act: the first law gave hedge fund managers a flat 4 percent tax rate if they moved their operations to the island, and the second offered investors complete tax exemption on dividends, interest, and capital gains if the investor lived on the island for half the year. Students from the University of Puerto Rico went on strike to protest the tuition hikes, closing ten of the eleven campuses of the UPR system. For a timeline

of events, see "Puerto Rico's Economic Crisis Timeline," Center for Puerto Rican Studies, Hunter College, accessed February 4, 2020, https://centropr.hunter.cuny.edu/sites/default/files/PDF_Publications/Puerto-Rico-Crisis-Timeline-2017.pdf. See also Ed Morales, "Puerto Rico's Unjust Debt," in Bonilla and LeBrón, *Aftershocks of Disaster*, 211–23.

116. Yaisha Vargas, "Puerto Rico Watches Corporate Tax Breaks Finally Expire," *Puerto Rico Herald*, May 25, 2005, http://www.puertorico-herald.org/issues2/2005/vol09n23/PRWatchCorpTax.html; Heriberto Martínez-Otero and Ian J. Seda-Irizarry, "The Origins of the Puerto Rico Debt Crisis," *Jacobin*, August 10, 2015, https://www.jacobinmag.com/2015/08/puerto-rico-debt-crisis-imf/; "Environmentalists Alarmed by Puerto Rico Polices," *MINA*, November 16, 2009, http://macedoniaonline.eu/content/view/9569/9/.

117. Jens Manuel Krogstad, "Historic Population Losses Continue across Puerto Rico," Pew Research Center, March 24, 2016, https://www.pewresearch.org/fact-tank/2016/03/24/historic-population-losses-continue-across-puerto-rico/. In a *New York Times* op-ed eighteen months before Hurricane Maria, Broadway actor and producer Lin-Manuel Miranda warned that the debt crisis would force professionals and working-class peoples to leave the island for higher-remunerated opportunities in the continental United States. "Lin-Manuel Miranda: Give Puerto Rico Its Chance to Thrive," *New York Times*, March 28, 2016, https://www.nytimes.com/2016/03/28/opinion/lin-manuel-miranda-give-puerto-rico-its-chance-to-thrive.html.

118. Gloria Guzman, "Household Income: 2016," *American Community Survey Briefs*, September 2017, https://www.census.gov/content/dam/Census/library/publications/2017/acs/acsbr16-02.pdf.

119. "HedgePapers No. 17—Hedge Fund Vultures in Puerto Rico," Hedge Clippers, July 10, 2015, http://hedgeclippers.org/hedgepapers-no-17-hedge-fund-billionaires-in-puerto-rico/.

120. Martínez-Otero and Seda-Irizarry, "Origins of the Puerto Rico Debt Crisis."

121. "Puerto Rico v. Franklin California Tax-Free Trust," *SCOTUSblog*, accessed March 19, 2020, https://www.scotusblog.com/case-files/cases/puerto-rico-v-franklin-california-tax-free-trust/; Lorraine McGowan, "First Circuit Rules Bankruptcy Code Preempts Puerto Rico's Recovery Act," *Orrick Law Firm Blog*, July 8, 2015, https://blogs.orrick.com/distressed-download/2015/07/08/first-circuit-rules-bankruptcy-code-preempts-puerto-ricos-recovery-act/.

122. Martínez-Otero and Seda-Irizarry, "Origins of the Puerto Rico Debt Crisis."

123. Klein, *Shock Doctrine*. According to Yarimar Bonilla and Marisol LeBrón, local residents call the board "La Junta," "which signals many Puerto Ricans' perception of the board as a dictatorial body that has seized power from the local government." See Bonilla and LeBrón, "Introduction," in *Aftershocks of Disaster*, 7.

124. The Federal Oversight and Management Board for Puerto Rico, Special Investigation Committee, *Independent Investigator's Final Investigative Report*, August 20, 2018, 2, https://assets.documentcloud.org/documents/4777926/FOMB-Final-Investigative-Report-Kobre-amp-Kim.pdf.

125. Melba Acosta Febo examined Puerto Rico's debt crisis and was much more optimistic about the economic growth PROMESA might generate despite Hurricane

Maria. See Melba I. Acosta Febo, "La reestructuración de la deuda como respuesta a la crisis fiscal de Puerto Rico y la recuperación luego del huracán María," *Revista Jurídica Universidad de Puerto Rico* 87, no. 3 (2018): 821–61. See also "Trump Says Puerto Rico in Trouble after Hurricane, Suffering from 'Broken Infrastructure' and 'Massive Debt,'" CNBC, September 26, 2017, https://www.cnbc.com/2017/09/25/trump-says-puerto-rico-in-trouble-after-hurricane-debt-must-be-dealt-with.html.

126. Luis E. Rodríguez Rivera discusses Puerto Rico's social and economic vulnerabilities in "La protección de derechos humanos como imperativo ante la coyuntura de desastres naturales y desigualdad socioeconómica: Mirada a Puerto Rico tras el huracán Maria," *Revista Jurídica Universidad de Puerto Rico* 87, no. 3 (2018): 772–88. See also "UN Experts Sound Alarm on Mounting Rights Concerns in Puerto Rico in Wake of Hurricane Maria," *UN News*, October 30, 2017, https://news.un.org/en/story/2017/10/569602-un-experts-sound-alarm-mounting-rights-concerns-puerto-rico-wake-hurricane.

127. During the second half of the seventeenth century, the Danish West India Company acquired St. Croix, St. Thomas, and St. John to serve as economic and military outposts. The Danish West Indies, as they were called, became a slave-based plantation economy that produced sugar, cotton, indigo, tobacco, and other tropical agricultural products for export. The United States first tried to purchase the islands in 1867, but the Senate failed to ratify the treaty. After securing the location for the Panama Canal, the McKinley and Roosevelt administrations resumed efforts to acquire these islands as strategic outposts to protect U.S. shipping, but it wasn't until 1917, the same year Congress passed the Jones-Shafroth Act, that they acquired the islands.

128. Austin, "Economic and Fiscal Conditions in the U.S. Virgin Islands," 4.

129. Austin, 7.

130. Austin, 4.

131. Jason Bram, "Federal Reserve Bank of New York, Economic Press Briefing: Puerto Rico and the U.S. Virgin Islands after Hurricanes Irma and Maria," Federal Reserve Bank of New York website, August 2018, https://www.newyorkfed.org/press/pressbriefings/puerto-rico-and-the-us-virgin-islands.

132. "Business Advantages: U.S. Jurisdiction," U.S. Virgin Islands Economic Development Authority, accessed July 17, 2019, https://www.usvieda.org/business-advantages.

133. Lisa Snowden-McCray, "After the Storm: The U.S. Virgin Islands Still Suffering," *Crisis*, Winter 2018, 7, accessed January 27, 2021, EBSCO.

134. "FEMA Issues Federal Action Plan for Puerto Rico Hurricane Georges Recovery," FEMA, February 23, 1999, https://www.fema.gov/news-release/1999/02/23/fema-issues-federal-action-plan-puerto-rico-hurricane-georges-recovery; Bob Sector and Marita Hernandez, "U.S. Aid Not Fast Enough, Islanders Say," *Los Angeles Times*, September 23, 1989, https://www.latimes.com/archives/la-xpm-1989-09-23-mn-600-story.html.

135. Juana Summers, "Trump Attacks San Juan Mayor over Hurricane Relief," CNN, September 30, 2017, https://www.cnn.com/2017/09/30/politics/trump-tweets-puerto-rico-mayor/index.html.

136. Donald J. Trump, Twitter, July 18, 2019, https://twitter.com/realdonaldtrump/status/1151867786236440576?lang=en.; Robert Farley, "Trump Misleads on Aid

to Puerto Rico," Factcheck.org, April 2, 2019, https://www.factcheck.org/2019/04/trump-misleads-on-aid-to-puerto-rico/. In the summer of 2018, during the height of another intense hurricane season, the Trump administration ordered that $10 million of FEMA's funding (approximately 1 percent of its operational budget) be redirected to Immigration and Customs Enforcement for the immigration detention facilities and deportation campaigns that were of primary concern to the president.

137. The administration barred Puerto Rico from paying its fifteen-dollars-an-hour minimum wage on federally funded projects, for example, and prohibited the use of any funds to conduct the much-needed repairs on the island's outdated electrical grid.

138. According to city planner Ivis Garcia, Puerto Rico has a history of informal construction: "Anywhere between 585,000–715,000 (45–55 percent) of homes and commercial buildings in Puerto Rico have been constructed without building permits or following land use codes, according to a 2018 study of the Puerto Rico Builders Association." An estimated 260,000 homes in Puerto Rico did not have titles or deeds. According to Garcia, FEMA's denials have proved "too hard for many Puerto Ricans to overcome because, though [deeds and titles] are not transactions that are costly, they are not part of the culture." Ivis Garcia, "The Lack of Proof of Ownership in Puerto Rico Is Crippling Repairs in the Aftermath of Hurricane Maria," *Human Rights* 44 (2019), Nexis Uni.

139. Elizabeth Delgado Santos et al. v. FEMA et al., United States District Court for the District of Massachusetts, accessed July 30, 2019, https://www.latinojustice.org/sites/default/files/First%20Amended%20Class%20Action%20Complaint%20Seeking%20Injunctive%20and%20Declaratory%20Relief%204.18%20cv%2040111%20TSH.pdf; "Transitional Assistance Program," FEMA, accessed July 30, 2019, https://www.fema.gov/transitional-shelter-assistance; letter from Elijah Cummings and Stacey Plaskett to William Long, U.S. House of Representatives, Committee on Oversight and Government Reform, August 14, 2018, https://oversight.house.gov/sites/democrats.oversight.house.gov/files/2018-08-14%20EEC%20Plaskett%20to%20Long-FEMA.pdf; "Senators Introduce Bill to Provide Stable Housing for Hurricane Survivors," Office of Senator Elizabeth Warren website, May 31, 2018, https://www.warren.senate.gov/newsroom/press-releases/senators-introduce-bill-to-provide-stable-housing-for-hurricane-survivors; "Warren Joins Espaillat in Leading Members of Congress to Urge Extension of Housing Assistance for Displaced Puerto Rican families," Office of Senator Elizabeth Warren website, June 28, 2018, https://www.warren.senate.gov/newsroom/press-releases/senators-introduce-bill-to-provide-stable-housing-for-hurricane-survivors. Espaillat reintroduced the bill in 2020 as the Housing Survivors of Major Disasters Act, Congress.gov, accessed August 11, 2021, https://www.congress.gov/bill/116th-congress/house-bill/2914.

140. Elizabeth Wolkomir, "How Is Food Assistance in Puerto Rico Different from the Rest of the United States?," Center on Budget and Policy Priorities, November 27, 2017, https://www.cbpp.org/research/food-assistance/how-is-food-assistance-different-in-puerto-rico-than-in-the-rest-of-the; Alice Thomas, "Keeping Faith with Our Fellow Americans: Meeting the Urgent Needs of Hurricane Maria Survivors in Puerto Rico," *Refugees International*, December 2017, 1, https://tinyurl.com/y7av3578. The nongovernmental organization urged the federal government to adopt in Puerto

Rico the "international best practices" endorsed by USAID's Office of Foreign Disaster Assistance.

141. Blanca Di Julio, Calley Muñana, and Maryanne Brodle, *Views and Experiences of Puerto Ricans One Year after Hurricane Maria*, Kaiser Family Foundation, September 12, 2018, https://www.kff.org/other/report/views-and-experiences-of-puerto-ricans-one-year-after-hurricane-maria/.

142. Some 40 percent of low-income families reported difficulty paying utilities; 38 percent had difficulty purchasing food; and 21 percent could not secure school materials for their children's education. Youth Development Institute of Puerto Rico, "The Impact of Hurricane Maria on Children in Puerto Rico," Youth Development Institute of Puerto Rico Google Doc, accessed March 27, 2021, https://drive.google.com/file/d/0B_GGLGhvE3QueDJYWTJNTThCTFE5YkExbnllQ3VOcEtKeWFF/view.

143. Greg Allen, "After 2 Hurricanes, a Floodgate of Mental Health Issues in the U.S. Virgin Islands," National Public Radio, April 23, 2019, https://www.npr.org/2019/04/23/716089187/after-two-hurricanes-a-floodgate-of-mental-health-issues-in-the-virgin-islands.

144. Bryan Lynn, "Puerto Rico Students Still Suffer the Effects of Hurricane Maria," Voice of America, October 2, 2018, https://learningenglish.voanews.com/a/puerto-rico-students-still-suffer-effects-of-hurricane-maria/4596814.html; Scott Jaschik, "Colleges Offer to Help Puerto Rican students," *Inside Higher Education*, October 16, 2017, https://www.insidehighered.com/quicktakes/2017/10/16/colleges-offer-help-puerto-rican-students.

145. Deirdre Walsh and Kevin Liptak, "Federal Response to Hurricane Maria Slowly Takes Shape," CNN, September 25, 2017, https://www.cnn.com/2017/09/25/politics/puerto-rico-hurricane-maria-aid-donald-trump/index.html.

146. Edwin Meléndez and Jennifer Hinojosa, "Estimates of Post–Hurricane Maria Exodus from Puerto Rico," Center for Puerto Rican Studies at Hunter College, October 2017, https://centropr.hunter.cuny.edu/sites/default/files/RB2017-01-POST-MARIA%20EXODUS_V3.pdf.

147. "State-to-State Migration Flows: 2018," U.S. Census Bureau, February 20, 2020, https://www.census.gov/data/tables/time-series/demo/geographic-mobility/state-to-state-migration.html.

148. "Nevada and Idaho Are the Nation's Fastest Growing States," U.S. Census Bureau, December 19, 2018, https://www.census.gov/newsroom/press-releases/2018/estimates-national-state.html.

149. Jason Schachter and Antonio Bruce, "Estimating Puerto Rico's Population after Hurricane Maria," U.S. Census Bureau, August 19, 2020, https://www.census.gov/library/stories/2020/08/estimating-puerto-rico-population-after-hurricane-maria.html; "Nevada and Idaho Are the Nation's Fastest Growing States."

150. The U.S. Census Bureau's American Community Survey and Puerto Rico Community Survey are good resources for estimating population movement, but the latter halted data collection during October–December 2017, and the former, which provides estimates annually, has a two-month residency requirement for inclusion in its surveys. Migration data collected from the American Community Survey tend to lag measurement of actual migration events. Monica Alexander, Kivan Polimis, and

Emilo Zagheni, "The Impact of Hurricane Maria on Out-Migration from Puerto Rico: Evidence from Facebook Data," *Population and Development Review* 45, no. 3 (September 2019), 617–30; Schachter and Bruce, "Estimating Puerto Rico's Population after Hurricane Maria."

151. "New Estimates: 135,000+ Post-Maria Puerto Ricans Relocated to Stateside," Center for Puerto Rican Studies, March 2018, https://centropr.hunter.cuny.edu/sites/default/files/data_sheets/PostMaria-NewEstimates-3-15-18.pdf.

152. Martín Echenique and Luis Melgar, "Mapping Puerto Rico's Hurricane Migration with Mobile Phone Data," *CityLab*, May 11, 2018, https://www.citylab.com/environment/2018/05/watch-puerto-ricos-hurricane-migration-via-mobile-phone-data/559889/. For other estimates see Jason Bram, "Federal Reserve Bank of New York, Economic Press Briefing: Puerto Rico and the U.S. Virgin Islands after Hurricanes Irma and Maria," Federal Reserve Bank of New York website, August 2018, https://www.newyorkfed.org/press/pressbriefings/puerto-rico-and-the-us-virgin-islands; and Antonio Flores and Jens Manuel Krogstad, "Puerto Rico's Population Declined Sharply after Hurricanes Maria and Irma," Pew Research Center, July 26, 2019, https://www.pewresearch.org/fact-tank/2019/07/26/puerto-rico-population-2018/.

153. Florida Division of Emergency Management, "Since October 3rd, more than 302,000 individuals have arrived in Florida from Puerto Rico through @FlyTPA @MCO and @iflymia," Twitter, January 4, 2018, https://tinyurl.com/y7egurgp; "Governor Scott Declares State of Emergency for Hurricane Maria to Support Puerto Rico," Florida Disaster.org, accessed August 5, 2019, https://www.floridadisaster.org/news-media/news/gov.-scott-declares-state-of-emergency-for-hurricane-maria-to-support-puerto-rico/.

154. Alfonso Chardy, "How Puerto Ricans Are Changing the Face of Florida," *Miami Herald*, January 22, 2017, https://www.miamiherald.com/news/local/immigration/article128081289.html; "Mapping Florida's Puerto Rico Population," *Puerto Rico Report*, April 25, 2019, https://www.puertoricoreport.com/mapping-floridas-puerto-rican-population/#.Xkw29lB7mjJ; Amy Martinez, "Puerto Rican Factor: Greener Pastures in Florida," *Florida Trend*, September 28, 2016, https://www.floridatrend.com/article/20708/puerto-rican-factor-greener-pastures-in-florida.

155. Alexander, Polimis, and Zagheni, "Impact of Hurricane Maria on Out-Migration from Puerto Rico," 624; Jack DeWaard, Janna E. Johnson, and Stephen D. Whitaker, "Out-migration from and Return Migration to Puerto Rico after Hurricane Maria: Evidence from Consumer Credit Panel," *Population and Environment* 42, no. 1 (September 2020): 28–42.

156. Flores and Krogstad, "Puerto Rico's Population Declined Sharply after Hurricanes Maria and Irma."

157. The U.S. Census Bureau reported a population of 106,405 in 2010. By 2017 the population stood at 104,751. In 2018 it was 104,680, and in 2019, 104,578. See "Table 4c. Population, Housing Units, Land Area, and Density by Island and Estate for the U.S. Virgin Islands: 2010," U.S. Census Bureau, accessed February 17, 2020, https://www.census.gov/population/www/cen2010/cph-t/cph-t-8.html; and "United States Virgin Islands Population 2019," World Population Review, accessed August 5, 2019, http://worldpopulationreview.com/countries/united-states-virgin-islands-population/.

158. Ricky Martin, "THIS IS NOT RIGHT. WE ARE IN THE MIDDLE OF A HUMANITARIAN CRISIS. THERE SHOULD BE LAWS AGAINST THIS," Twitter, September 25, 2017, https://twitter.com/ricky_martin/status/912357893655031808?lang=en. See also Chabeli Herrera, "Is a $786 One-Way Flight Out of Puerto Rico Price Gouging? Airlines Say No," *Bradenton Herald*, September 27, 2017, https://www.bradenton.com/article175607971.html.

159. James Hohmann, "The Daily 202: Trump's Katrina? Influx of Puerto Ricans after Hurricane Maria Could Tip Florida towards Democrats," *Washington Post*, September 28, 2017, https://www.washingtonpost.com/news/powerpost/paloma/daily-202/2017/09/28/daily-202-trump-s-katrina-influx-of-puerto-ricans-after-hurricane-maria-could-tip-florida-toward-democrats/59cc037c30fb0468cea81c2f/.

160. Antonio Flores, Mark Hugo Lopez, and Jens Manuel Krogstad, "Hispanic Voter Registration Rises in Florida, but Role of Puerto Ricans Remains Unclear," Pew Research Center, October 12, 2018, https://www.pewresearch.org/fact-tank/2018/10/12/hispanic-voter-registration-rises-in-florida-but-role-of-puerto-ricans-remains-unclear/; Carmen Sesin, "Trump Cultivated the Latino Vote in Florida, and It Paid Off," *NBC News*, November 4, 2020, https://www.nbcnews.com/news/latino/trump-cultivated-latino-vote-florida-it-paid-n1246226; "Puerto Rico Statehood Referendum (2020)," Ballotpedia, accessed November 11, 2021, https://ballotpedia.org/Puerto_Rico_Statehood_Referendum_(2020).

161. By comparison, two years after Hurricane Harvey, 3,700 projects had been funded in Texas and 3,700 in Florida. Mark Walker and Zolan Kanno-Youngs, "FEMA's Hurricane Aid to Puerto Rico and Virgin Islands Has Stalled," *New York Times*, November 27, 2019, https://www.nytimes.com/2019/11/27/us/politics/fema-hurricane-aid-puerto-rico-virgin-islands.html.

162. In an April 2019 interview on MSNBC, White House Deputy Press Secretary Hogan Gidley referred to Puerto Rico as "that country," reinforcing the perception that the administration did not view the territories as American. Rebecca Morin, "White House Aide Calls Puerto Rico 'That Country,' Says It Was 'Slip of the Tongue,'" Politico, April 2, 2019, https://www.politico.com/story/2019/04/02/gidley-puerto-rico-country-1249076; Aaron Blake, "Trump Keeps Talking about Puerto Rico Like It Isn't Part of the U.S. It Doesn't Seem Like a Mistake," *Washington Post*, April 2, 2019, https://www.washingtonpost.com/politics/2019/04/02/white-houses-seemingly-deliberate-effort-otherize-puerto-rico/?utm_term=.5540973a522e.

163. Katie Galioto, "Trump Boasts He's Taken Better Care of Puerto Rico 'Than Any Living Human Being,'" Politico, March 28, 2019, https://www.politico.com/story/2019/03/28/trump-boats-hes-taken-better-care-of-puerto-rico-than-any-living-human-being-1243077.

164. Alexia Fernández Campbell and Umair Irfan, "Puerto Rico's Deal with Whitefish Was Shady as Hell, New Records Show," Vox, November 15, 2017, https://www.vox.com/policy-and-politics/2017/11/15/16648924/puerto-rico-whitefish-contract-congress-investigation. PREPA was "once considered a crowning achievement of effective public management [because it brought] electrification to remote parts of the island that private companies had neglected to serve" but over time was accused of mismanagement. In 2015, for example, the utility was sued for overcharging

customers for low-quality oil for over a decade. "Puerto Rico Electric and Power Authority (PREPA)," Hagens Berman Law Firm, accessed February 24, 2020, https://www.hbsslaw.com/cases/puerto-rico-electric-power-authority-prepa; Cathy Kunkel, "Rosselló Steps Down after Puerto Rico Rises," *Jacobin*, July 26, 2019, https://jacobinmag.com/2019/07/puerto-rico-protest-ricardo-rossello.

165. "Former Secretary of Puerto Rico Department of Education and Former Executive Director of Puerto Rico Health Insurance Administration Indicted with Four Others for Conspiracy, Wire Fraud, Theft of Government Funds, and Money Laundering," U.S. Department of Justice, U.S. Attorney's Office, District of Puerto Rico, July 10, 2019, https://www.justice.gov/usao-pr/pr/former-secretary-puerto-rico-department-education-and-former-executive-director-puerto-1; Patricia Mazzei, "Puerto Rico Ex Officials Accused of Steering $15.5 Million in Contracts to Consultants," *New York Times*, July 10, 2019, https://www.nytimes.com/2019/07/10/us/puerto-rico-corruption.html.

166. For its work, Centro de Periodismo Investigativo received the 2020 Louis M. Lyons Award for Conscience and Integrity in Journalism. See "Louis M. Lyons Award for Conscience and Integrity in Journalism," Nieman Foundation, accessed February 24, 2020, https://nieman.harvard.edu/awards/louis-lyons-award/. See also Fernando Tormos-Aponte, "How an Investigative Journalism Center Helped Oust Puerto Rican Governor Rosselló," *In These Times*, August 1, 2019, https://inthesetimes.com/article/21989/investigative-journalism-rossello-puerto-rico-protests; and Ray Sánchez, "These Are Some of the Leaked Chat Messages at the Center of Puerto Rico's Political Crisis," CNN, July 17, 2019, https://www.cnn.com/2019/07/16/us/puerto-rico-governor-rossello-private-chats/index.html.

167. Mike Wall, "Earthquakes in Puerto Rico Have Changed the Landscape. Satellites Can See It from Space," Space.com, January 13, 2020, https://www.space.com/puerto-rico-earthquake-effects-satellite-views.html.

168. The last tsunami to affect the Virgin Islands occurred on November 18, 1867, when a magnitude 7.5 earthquake affected the Anegada trough, located between St. Croix and St. Thomas. "The 1867 Virgin Island Tsunami," University of Southern California Tsunami Research Center, accessed January 21, 2021, http://www.tsunamiresearchcenter.com/tsunami-archive/the-1867-virgin-island-tsunami/; "U.S. Virgin Islands on High Alert after 9 Earthquakes Strike Puerto Rico in One Day," *Virgin Islands Consortium*, January 7, 2020, https://viconsortium.com/vi-top_stories/virgin-islands-u-s-virgin-islands-on-high-alert-after-9-earthquakes-strike-puerto-rico-in-one-day-state-of-emergency-declared-in-puerto-rico.

169. "2017 Hurricane Season FEMA After-Action Report," ii–iii, 49–50.

170. Luis A. Avilés, "Congress Must Create an Energy Policy Specific to Its Tropical Island Territories," *Microjuris*, November 2, 2017, Nexis Uni.

Conclusion: Moving Forward

1. New Zealand's Immigration and Protection Tribunal is "an independent body that hears appeals from decisions made by Immigration New Zealand, including those of refugee and protection officers within the Refugee Status Branch." The tribunal

was preceded by the New Zealand Refugee Status Appeals Authority, which in 2000 overheard eight other cases involving claimants from the island state of Tuvalu who sought recognition as refugees in New Zealand "due to various environmental factors, including 'inundation, coastal erosion, salination of the water table, combined with factors at the individual and household levels.'" "New Zealand: 'Climate Change Refugee' Case Overview," Library of Congress, accessed October 20, 2017, https://www.loc.gov/item/2016295703/.

2. "New Zealand: 'Climate Change Refugee' Case Overview."

3. United Nations International Covenant on Civil and Political Rights, "Views Adopted by the Committee under Article 5 (4) of the Optional Protocol," *Columbia University Law School Blog*, accessed November 7, 2021, http://blogs2.law.columbia.edu/climate-change-litigation/wp-content/uploads/sites/16/non-us-case-documents/2020/20200107_CCPRC127D27282016-_opinion.pdf.

4. "Central America," Center for Climate and Security, April 17, 2019, https://climateandsecurity.org/tag/central-america/; "Secretary Nielsen Signs Historic Regional Compact with Central America to Stem Irregular Migration at the Source, Confront U.S. Border Crisis," U.S. Department of Homeland Security, March 28, 2019, https://www.dhs.gov/news/2019/03/28/secretary-nielsen-signs-historic-regional-compact-central-america-stem-irregular.

5. Koko Warner, Charles Ehrhart, Alex de Sherbinin, Susana Adamo, and Tricia Chai-Onn, *In Search of Shelter: Mapping the Effects of Climate Change on Human Migration and Displacement*, Center for International Earth Science Information Network, May 2009, 1, http://www.ciesin.columbia.edu/documents/clim-migr-report-june09_media.pdf.

6. Some 234 scientists analyzed the data on climate change. The Intergovernmental Panel on Climate Change, *Sixth Assessment Report*, IPCC, August 2021, https://www.ipcc.ch/.

7. "Secretary-General Calls Latest IPCC Report 'Code Red for Humanity,' Stressing 'Irrefutable Evidence of Human Influence,'" United Nations, August 9, 2021, https://www.un.org/press/en/2021/sgsm20847.doc.htm.

8. Rabab Fatima, Anita Jawadurovna, and Sabira Coelho, "Human Rights, Climate Change, Environmental Degradation and Migration: A New Paradigm," Migration Policy Institute, March 2014, 6, http://www.migrationpolicy.org/research/human-rights-climate-change-environmental-degradation-and-migration-new-paradigm; Kanta Kumari Rigaud, Alex de Sherbinin, Bryan Jones, Jonas Bergmann, Viviane Clement, Kayly Ober, Jacob Schewe, et al., *Groundswell: Preparing for Internal Climate Migration*, World Bank, 2018, https://openknowledge.worldbank.org/handle/10986/29461; "What Climate Change Means for Asia, Africa, and the Coastal Poor," World Bank, June 19, 2013, http://www.worldbank.org/en/news/feature/2013/06/19/what-climate-change-means-africa-asia-coastal-poor; "Addressing Climate Change and Migration in Asia and the Pacific," Asian Development Bank, March 2012, https://www.adb.org/publications/addressing-climate-change-and-migration-asia-and-pacific; Carolina Fritz, "Climate Change and Migration: Sorting through Complex Issues without the Hype," *Migration Information Source*, March 4, 2010, https://www.migrationpolicy.org/article/climate-change-and-migration-sorting-through-complex-issues-without-hype.

9. Sociologist Saskia Sassen defines "development malpractice" as "international economic policies ... that have undermined local economies in the Global South, expanded poverty, and forced people into the migrant stream." Saskia Sassen, "Three Emergent Migrations: An Epochal Change," *Sur: International Journal on Human Rights*, no. 23 (July 2016): 29–41, https://sur.conectas.org/wp-content/uploads/2016/09/2-sur-23-ingles-saskia-sassen.pdf.

10. Kathleen Newland, *Climate Change and Migration Dynamics*, Migration Policy Institute, September 2011, 1–3, http://www.migrationpolicy.org/research/climate-change-and-migration-dynamics.

11. See, for example, Parag Mahajan and Dean Yang, "Taken by Storm: Hurricanes, Migrant Networks, and U.S. Immigration," National Bureau of Economic Research, Working Paper No. w23756, August 2017, http://www.nber.org/papers/w23756; Dean Yang and Parag Mahajan, "Hurricanes Drive Immigration to the United States," *The Conversation*, September 15, 2017, https://theconversation.com/hurricanes-drive-immigration-to-the-us-83755.

12. Abrahm Lustgarten, "Where Will Everyone Go?," *ProPublica*, July 23, 2020, https://features.propublica.org/climate-migration/model-how-climate-refugees-move-across-continents/.

13. Nick Watts et al., "The Lancet Countdown on Health and Climate Change: From 25 Years of Inaction to a Global Transformation for Public Health," *The Lancet*, October 30, 2017, http://www.thelancet.com/journals/lancet/article/PIIS0140-6736(17)32464-9/fulltext; Charles Geisler and Ben Currens, "Impediments to Inland Resettlement under Conditions of Accelerated Sea Level Rise," *Land Use Policy* 66 (2017): 322–30, https://doi.org/10.1016/j.landusepol.2017.03.029; Rigaud et al., *Groundswell*; "Land and Human Security," United Nations Convention to Combat Desertification, accessed August 22, 2017, http://www2.unccd.int/issues/land-and-human-security; Ilmi Granoff, Jason Eis, Will McFarland, Chris Hoy, et al., *Zero Poverty, Zero Emissions: Eradicating Extreme Poverty in the Climate Crisis* (London: Overseas Development Institute, September 2015), 3, https://odi.org/en/publications/zero-poverty-zero-emissions-eradicating-extreme-poverty-in-the-climate-crisis/.

14. Newland, *Climate Change and Migration Dynamics*, 2–3; Viviane Clement, Kanta Kumari Rigaud, Alex de Sherbinin, Bryan Jones, Susana Adamo, Jacob Schewe, Nian Sadiq, and Elham Shabahat, *Groundswell Part 2: Acting on Internal Climate Migration*, World Bank, 2021, https://openknowledge.worldbank.org/handle/10986/36248.

15. Migrants transform landscapes not only through the depletion of resources and the introduction of new biological species and disease but also by demanding legal rights and privileges in their new locales. For a discussion of the ways migrants have physically transformed landscapes over time, see Marco Armiero and Richard Tucker, eds., *Environmental History of Modern Migrations* (New York: Routledge, 2017). Several of the essays in this anthology examine the ways migrants have brought deforestation, soil depletion, and reduction of biodiversity.

16. In their important study of armed conflicts from 1980 to 2010, Carl-Friedrich Schleussner and his colleagues found "evidence in global data sets that risk of armed-conflict outbreak is enhanced by climate-related disaster occurrence in ethnically fractionalized countries. . . . [But while] we find no indications that environmental

disasters directly trigger armed conflicts, our results imply that disasters might act as a threat multiplier in several of the world's most conflict prone regions." Carl-Friedrich Schleussner, Jonathan F. Donges, Reik V. Donner, and Hans Joachim Schellnhuber, "Armed-Conflict Risks Enhanced by Climate-Related Disasters in Ethnically Fractionalized Countries," *Proceedings of the National Academy of Sciences* 113, no. 33 (2016): 9216–21, published ahead of print July 25, 2016, doi:10.1073/pnas.160161111. See also Katharina Nett and Lukas Rüttinger, *Insurgency, Terrorism and Organised Crime in a Warming Climate: Analysing the Links between Climate Change and Non-state Armed Groups* (Berlin: Adelphi, October 2016), iii–iv, https://www.adelphi.de/en/publication/insurgency-terrorism-and-organised-crime-warming-climate. Another study found that the probability of civil instability in countries in the tropics doubled in the years of El Niño–Southern Oscillation. The authors concluded that this weather phenomenon played a contributing factor in 21 percent of all civil conflicts from 1950 to 2004. See also Solomon M. Hsiang et al., "Civil Conflicts Are Associated with the Global Climate," *Nature* 476, no. 7361 (August 25, 2011): 438–41. Since the 1980s, the Pacific Institute has tracked international conflicts over water and reported 94 incidents of water conflict between 2000 and 2010 and 129 incidents between 2011 and 2017. The Pacific Institute, "The World's Water," Worldwater.org, accessed August 27, 2017, http://worldwater.org/water-conflict/. Other scholars of security have examined how climate change may be a contributing factor in the rise and growth of non-state-armed groups, many of which then use food, fuel, and natural resources as targets or as tools of strategy. See, for example, Elizabeth Chalecki, "A New Vigilance: Identifying and Reducing the Risks of Environmental Terrorism," Woodrow Wilson Center for International Scholars, September 26, 2001, http://pacinst.org/publication/a-new-vigilance-identifying-and-reducing-risk-of-environmental-terrorism/; and Nett and Rüttinger, *Insurgency, Terrorism and Organised Crime in a Warming Climate*. Janet L. Sawin of the Worldwatch Institute warned that if governments do not meet the basic needs of populations during periods of drought and famine, affected populations might turn to extralegal groups that promise to meet their needs. See Janet L. Sawin, "Global Security Brief #3: Climate Change Poses Greater Security Threat Than Terrorism," Worldwatch Institute, accessed August 27, 2017, http://www.worldwatch.org/node/77.

17. UN Environment Programme, "Protecting the Environment during Armed Conflict: An Inventory and Analysis of International Law," United Nations Environment Programme website, November 2009, http://www.un.org/zh/events/environmentconflictday/pdfs/int_law.pdf.

18. One 2018 Pentagon study presented to Congress revealed that roughly half of the 3,500 military installations worldwide have been affected by drought, wildfires, extreme wind conditions, extreme temperatures, or storm surge–related flooding. U.S. Department of Defense, *Climate-Related Risk to DoD Infrastructure Initial Vulnerability Assessment Survey*, January 2018, https://climateandsecurity.files.wordpress.com/2018/01/tab-b-slvas-report-1-24-2018.pdf; "Hampton Roads," Center for Climate and Security, accessed May 21, 2020, https://climateandsecurity.org/tag/hampton-roads/; "Hampton Roads Climate Change Adaptation Project," Georgetown Climate Center Adaptation Clearinghouse, accessed May 21, 2020, https://www

.adaptationclearinghouse.org/resources/hampton-roads-climate-change-adaptation-project.html.

19. One example of the role of armed forces in humanitarian intervention: in the year 2000, the country of Mozambique was hit by consecutive cyclones and historic flooding affecting 2 million people. Eleven nations sent personnel to assist with rescue and evacuation efforts, including the United States, which contributed personnel and logistics for these humanitarian efforts. This disaster occurred just eight years after the peace accords ended the sixteen-year civil war in Mozambique. See Stockholm International Peace Research Institute, "Annex A: Floods and Cyclones in Mozambique," in *The Effectiveness of Foreign Military Assets in Natural Disaster Response* (Stockholm: SIPRI, 2008), 55–68, https://reliefweb.int/sites/reliefweb.int/files/resources/236476AD3257088DC125741000474F20-sipri_mar2008.pdf. More recently, the United States provided aid in 2019, after Cyclone Idai caused additional devastation. Kent Miller, "U.S. Military Supporting Disaster Relief in Mozambique," *Air Force Times*, accessed May 21, 2020, https://www.airforcetimes.com/news/your-air-force/2019/03/28/us-military-supporting-disaster-relief-efforts-in-mozambique/.

20. The drought and famine that affected Ethiopia in the 1980s, for example, exacerbated the instability caused by the ongoing civil war and contributed to the deaths of an estimated 1 million people and the out-migration of more than 600,000 refugees. Laura Hammond, "Ethiopians Who Survived the Famine: A Repatriation Success Story," *Migration Information Source*, July 1, 2005, http://www.migrationpolicy.org/article/ethiopians-who-survived-famine-repatriation-success-story. See also U.S. Department of Defense, *Quadrennial Defense Review Report*, February 2010, https://www.defense.gov/Portals/1/features/defenseReviews/QDR/QDR_as_of_29JAN10_1600.pdf; Achim Steiner, untitled report on the United National Environment Programme and NATO, North Atlantic Treaty Organization website, accessed November 7, 2021, https://www.nato.int/strategic-concept/pdf/UNEP.pdf; Mark Hand, "Trump Is Jeopardizing Pentagon's Efforts to Fight Climate Change, Retired Military Leaders Fear," *Think Progress*, July 13, 2017, https://thinkprogress.org/former-defense-officials-see-climate-action-lagging-e550994d65/; "Hagel to Address 'Threat Multiplier' of Climate Change," U.S. Department of Defense, October 13, 2014, https://www.defense.gov/News/Article/Article/603440/; Amar Causevic, "Facing an Unpredictable Threat: Is NATO Ideally Placed to Manage Climate Change as a Non-traditional Threat Multiplier?," *Connections: The Quarterly Journal* 16, no. 2 (2017): 59–80, http://connections-qj.org/article/facing-unpredictable-threat-nato-ideally-placed-manage-climate-change-non-traditional-threat?lipi=urn%3Ali%3Apage%3Ad_flagship3_profile_view_base%3B9Mzl%2F%2BFTR9uJ7WKe%2Fo2VzA%3D%3D; and National Intelligence Council, "Implications for U.S. National Security of Anticipated Climate Change," Office of the Director of National Intelligence website, September 21, 2016, 12–13, https://www.dni.gov/files/documents/Newsroom/Reports%20and%20Pubs/Implications_for_US_National_Security_of_Anticipated_Climate_Change.pdf.

21. Christian Parenti, *The Tropic of Chaos: Climate Change and the New Geography of Violence* (New York: Nation Books, 2011).

22. Nick Buxton and Ben Hayes, "Conclusion: Finding Security in a Climate-Changed World," in *The Secure and the Dispossessed: How the Military and the Corporations Are*

Changing a Climate-Changed World, ed. Nick Buxton and Ben Hayes (London: Pluto Press, 2015), 232.

23. Daniel Aldana Cohen, "New York Mag's Climate Disaster Porn Gets It Painfully Wrong," *Jacobin*, July 10, 2017, https://jacobinmag.com/2017/07/climate-change-new-york-magazine-response.

24. Rebecca Solnit, *A Paradise Built in Hell: The Extraordinary Communities That Arise in Disaster* (New York: Penguin Books, 2010).

25. Newland, *Climate Change and Migration Dynamics*, 2.

26. World People's Conference on Climate Change and the Rights of Mother Earth, "Universal Declaration of Rights of Mother Earth" (the "Cochabamba Declaration"), Cochabamba, Bolivia, April 24, 2010, https://pwccc.wordpress.com/2010/04/24/peoples-agreement/.

27. Sarah Jaquette Ray offers some useful tools for addressing climate anxiety in her book *A Field Guide to Climate Anxiety: How to Keep Your Cool on a Warming Planet* (Berkeley: University of California Press, 2020).

28. Margaret Besheer, "Volcano on St. Vincent Still Erupting," Voice of America, April 20, 2021, https://www.voanews.com/americas/volcano-st-vincent-still-erupting.

29. The humanitarian parole authority was first included in the 1952 Immigration Act (more popularly known as the McCarran-Walter Act). An Act to Revise the Laws Relating to Immigration, Naturalization, and Nationality, Public Law 82-414, *U.S. Statutes at Large* 66 (1952): 163–282; see also "§ Sec. 212.5 Parole of Aliens into the United States," National Archives, accessed October 19, 2017, https://www.uscis.gov/ilink/docView/SLB/HTML/SLB/0-0-0-1/0-0-0-11261/0-0-0-15905/0-0-0-16404.html.

30. The most well-known of these adjustment acts is the 1966 Cuban Adjustment Act. See "The Cuban Adjustment Act Cases," USCIS, accessed November 30, 2017, https://www.uscis.gov/ilink/docView/AFM/HTML/AFM/0-0-0-1/0-0-0-8624/0-0-0-10170.html.

31. "S. 2565—116th Congress: A Bill to Establish a Global Climate Change Resilience Strategy, to Authorize the Admission of Climate-Displaced Persons, and for Other Purposes," GovTrack.us, accessed July 27, 2020, https://www.govtrack.us/congress/bills/116/s2565. President Biden's executive order stated that the report, to be issued within 180 days, should include "at a minimum, discussion of the international security implications of climate-related migration; options for protection and resettlement of individuals displaced directly or indirectly from climate change; mechanisms for identifying such individuals, including through referrals; proposals for how these findings should affect use of United States foreign assistance to mitigate the negative impacts of climate change; and opportunities to work collaboratively with other countries, international organizations and bodies, non-governmental organizations, and localities to respond to migration resulting directly or indirectly from climate change." "Executive Order on Rebuilding and Enhancing Programs to Resettle Refugees and Planning for the Impact of Climate Change on Migration," White House, February 4, 2021, https://www.whitehouse.gov/briefing-room/presidential-actions/2021/02/04/executive-order-on-rebuilding-and-enhancing-programs-to-resettle-refugees-and-planning-for-the-impact-of-climate-change-on-migration/.

32. "When It Comes to Sea Level Rise, Not Every State Is the Same," SeaLevelRise.org, accessed July 23, 2020, https://sealevelrise.org/states/.

33. Henry Gass, "Tactical Retreat? As Seas Rise, Louisiana Faces Hard Choices," *Christian Science Monitor*, August 2, 2017, https://www.csmonitor.com/Environment/Inhabit/2017/0802/Tactical-retreat-As-seas-rise-Louisiana-faces-hard-choices; Henry Gass, "In Race against Rising Seas, Louisiana Scrambles to Save Dwindling Coast," *Christian Science Monitor*, August 3, 2017, https://www.csmonitor.com/Environment/Inhabit/2017/0803/In-race-against-rising-seas-Louisiana-scrambles-to-save-dwindling-coast. According to Gass, "The plan, which is reviewed every five years, analyzes the latest scientific research to predict how the coastline may change over the next five decades, then recommends various projects across the state to help protect and mitigate the effects of the changes."

34. Other studies have warned that a three- to six-foot rise in sea levels by the end of the century could leave 4–13 million Americans in coastal communities homeless. "When Rising Seas Hit Home: Hard Choices Ahead for Hundreds of U.S. Coastal Communities," Union of Concerned Scientists, July 2017, 2, http://www.ucsusa.org/sites/default/files/attach/2017/07/when-rising-seas-hit-home-full-report.pdf; Mathew E. Hauer, Jason M. Evans, and Deepak R. Mishra, "Millions Projected to Be at Risk from Moderate Sea-Level Rise in the Continental United States," *Nature Climate Change*, March 14, 2016, https://www.nature.com/articles/nclimate2961.epdf.

35. Brianna Fernandez, "Massive Hurricanes Could Change State and Local Populations," American Action Forum, November 30, 2017, https://www.americanactionforum.org/research/massive-hurricanes-change-state-local-populations/. The population of New Orleans fell from 484,674 (April 2000) to an estimated 230,172 after Katrina (July 2006); by July 2012, the population was back up to 369,250—76 percent of what it was in 2000. Allison Plyer, "Facts for Features: Katrina Impact," The Data Center, August 26, 2016, https://www.datacenterresearch.org/data-resources/katrina/facts-for-impact/; U.S. Department of Homeland Security, *The Federal Response to Hurricane Katrina: Lessons Learned*, February 2006, https://www.hsdl.org/?view&did=460536.

36. In 2016, Isle de Jean Charles became the first community in the United States to receive federal assistance for a large-scale retreat from the impacts of climate change. The Obama administration's Housing and Urban Development Office created the National Disaster Resilience Competition to fund innovative projects that help communities and states recover from disaster and reduce future risks. Louisiana received $92.6 million from the competition, of which $48 million was to go to relocating the people of Isle de Jean Charles. According to one Louisiana official, the test case allows the government to "learn as much as we can so other communities can do it more successfully in the future, less expensively in the future, and more efficiently in the future." Such efforts might help other Indigenous populations in the Mississippi Delta, such as the Grand Caillou/Dulac Band of the Biloxi-Chitimacha-Choctaw and the Grand Bayou Atakapa tribe, adapt to the realities of climate change. Tristan Baurick, "Houma Sugar Farms Are Finalists for Isle de Jean Charles Resettlement," NOLA.com, July 18, 2017, http://www.nola.com/environment/index.ssf/2017/07/houma_sugar_plantations_eyed f.html; "Tribal Resettlement," Isle de Jean Charles website, accessed July 23, 2020, http://www.isledejeancharles.com/our-resettlement; Kyle Mandel, "America's Climate Refugees Have Been Abandoned by Trump," *Mother Jones*, October 17, 2017, https://www.motherjones.com/environment/2017/10/climate-refugees-trump-hud/;

Neha Thirani Bagri, "The U.S. Is Relocating an Entire Town Because of Climate Change. And This Is Just the Beginning," Quartz, June 5, 2017, https://qz.com/994459/the-us-is-relocating-an-entire-town-because-of-climate-change-and-this-is-just-the-beginning/; Coral Davenport and Campbell Robertson, "Resettling the First American 'Climate Refugees,'" New York Times, May 3, 2016, A1, A11.

37. General Accounting Office, *Alaska Native Villages: Limited Progress Has Been Made on Relocating Villages Threatened by Flooding and Erosion*, June 2009, 1, http://www.gao.gov/new.items/d09551.pdf; U.S. Army Corps of Engineers, "Relocation Planning Project Master Plan: Kivalina, Alaska," Adaptation Clearinghouse, Georgetown Climate Center, June 2006, 1, https://www.adaptationclearinghouse.org/resources/usace-kivalina-alaska-relocation-master-plan.html.

38. Bagri, "U.S. Is Relocating an Entire Town Because of Climate Change."

39. The Obama administration, for example, gave "climate resilience grants" totaling $1 billion to thirteen states to supplement state funding of infrastructure projects. Elizabeth Kolbert, "The Climate of Man, Part 1," *New Yorker*, April 25, 2005, https://www.newyorker.com/magazine/2005/04/25/the-climate-of-man-i; Fritz, "Climate Change and Migration."

40. NOAA National Centers for Environmental Information, "U.S. Billion-Dollar Weather and Climate Disasters," National Oceanic and Atmospheric Administration website, 2021, https://www.ncdc.noaa.gov/billions/.

INDEX

ABC agreement (1990), 72, 187n119
Agricultural Modernization Law (1992), 186n104
agricultural production: climate change's effects on, 48, 87, 152n3; in Honduras, 55, 56, 64, 77, 78, 180n42; in Montserrat, 43, 168n24; in Puerto Rico and USVI, 98, 105, 118, 124, 125
Agua Zarca dam, 88
Aguilar, Roberto, 52
AIR (Associated Independent Recording) Studio, 21, 166n7
airport development, 41–42
Alaska, 114, 148
Alemán, Arnoldo, 66, 185n101
American Baptist Church v. Thornburgh, 72, 187n119
American Dream and Promise Act (2021), 80, 194n186
American Recovery and Reinvestment Act (2009), 122
Andrés, José, 105
Anguilla, 29, 166n9
Antigua, 14, 26, 29, 32, 35, 41, 145, 169n46, 174n104, 197n2

archival heritage, loss of, 20, 165n1, 166n10
Arzu, Álvaro, 75
Atwood, Brian, 61, 65, 70, 182n66
avalanche, 51. *See also* landslides
Ávila-Marrero, Angela, 134
Avilés, Luis A., 136
Azorean Refugees Act (1958), 9, 10, 162n45
Azores, 9, 10, 161n41, 162n45

banana industry, 55, 77, 98. *See also* agricultural production
Bangladesh, 157n26
bankruptcy, 123. *See also* public debt
Barbados, 20, 145, 166n6
Barbuda, 14, 26, 29, 32, 45, 93, 197n2
Begnaud, David, 103–4
Behrman, Simon, 6
Berger, Oscar, 12
Berger, Samuel "Sandy," 58
Biden administration, 84, 128, 147, 222n31
Blair, Tony, 28
Bohoslavsky, Juan Pablo, 124
Bonilla, Yarimar, 117, 123
Bossert, Tom, 96
Brandt, David, 26, 30, 37–38

INDEX

British citizenship, 29, 32, 45
British Overseas Territories, 29, 166n9. *See also* Montserrat
British Overseas Territory Act (2002), 21
Buchanan, Jeffrey, 116
Bush (G. H. W.) administration, 127
Bush (G. W.) administration, 12, 15, 34, 36, 37, 45, 65, 82–83, 163n52
Buxton, Nick, 143

Cáceres, Berta, 88, 89
Calderon Sol, Armando, 75
California, 73, 76, 77, 84, 108, 192n167
Canada, 14, 32, 49, 69, 77, 82, 142
Caribbean Basin Initiative, 64, 184n88
Caribbean Development Bank, 40
CARICOM, 27, 33, 40, 41
Cartagena Declaration (1984), 5
Casa Pueblo, Puerto Rico, 105, 204n54
Casillas, Rodolfo, 90
casualties: of children, disaster-related, 3, 25, 86; in Colombia, 183n72; in El Salvador, 193n179; in Ethiopia, 221n20; in Guatemala, 177n9; in Honduras and Nicaragua, 15, 56, 86, 88–89; in Montserrat, 25, 29; in Puerto Rico and USVI, 98–99, 103, 104, 111, 126, 134, 200n23. *See also* wars
"catastrophic natural calamity," as term, 9, 10
celebrity fundraising and disaster aid, 105–6, 134, 204n57
Center for Puerto Rican Studies, 130, 131
Center for the Investigation and Promotion of Human Rights, 57
Centers for Disease Control (CDC), 101
Central American Commission on Environment and Development, 85, 87
Central American Emergency Trust Fund, 63
Central American Free Trade Agreement, 65
Central American Integration System, 81, 192n170
Central American Volcanic Arc, 48
Centro de Periodismo Investigativo, 102, 134

Chao, Elaine, 113
Chesney, Kenny, 105–6
children: disaster-related casualties of, 3, 25, 86; effects of hurricanes in Puerto Rico and USVI on, 129–30; exploitation of, 86; post-traumatic stress of, 39–40; response to loss of parents, 52; vulnerability of, 3
Chiquita Brands, 31, 55
cholera, 52, 59
Christen-Christensen, Donna, 37
Christian Aid, 55
Citibank, 64
citizen activism: in Honduras, 68, 87–88; in Puerto Rico and USVI, 103–6, 203n42
citizenship: of Montserratians, 29, 32, 45; of Puerto Ricans and Virgin Islanders, 16, 102, 110, 114–30; UNHCR on, 8. *See also* legal definition of "refugee"; TPS (Temporary Protected Status); voting rights
Civil Authority Information Systems, 104
climate change, 1–2, 151n2; Biden's executive order on, 147, 222n31; cases on historical responses to, 2–5; effects on agricultural production, 30, 85, 87, 143, 152n3; effects on Central American isthmus, 47–48, 84–85, 86–87, 90, 145; future of migration due to, 140–49, 153n7, 158n30; Montserrat and, 44–46; UN's response to, 141–43, 159n35
climate, defined, 151n1
climate migrant, as term, 6
climate refugees, as term, 5–6, 148, 155n19
Clinton, Hillary Rodham: on climate change, 90; on Honduran and Nicaraguan aid, 53, 59, 60–61; on migrant repatriation, 76; in Puerto Rico aid, 113, 116; support of Zelaya coup by, 67, 185n100
Clinton administration: Central American disaster aid and, 58, 60–67, 185n97; on disaster-driven migration from Central America, 70–71; migrant detentions and, 76–77; on Montserrat disaster, 33; TPS and deportations by, 82, 83
coastal erosion, 43, 147–48, 223n34. *See also* landslides; soil erosion

226

INDEX

Cochabamba Declaration, 144
coffee industry, 56, 98, 105. *See also* agricultural production
Cohen, Daniel Aldana, 143–44
Colom, Álvaro, 12
Colombia, 61, 183n72
coloniality and citizenship, 116–30
coloniality of disaster, as phrase, 117
Columbus, Christopher, 20
La Comisión para el Esclarecimineto Histórico report (Guatemala), 67
Committee of Families of the Detained and the Disappeared of Honduras, 88, 89
Commonwealth Nutrition Assistance Program (NAP), 129
commonwealth status and disaster aid in Montserrat, 14, 32, 45, 144–45. *See also* Montserrat; territorial status and disaster aid in Puerto Rico and USVI
communication, post-disaster, 24, 26, 43, 102–7, 110, 127, 131, 205n62. *See also* radio stations
compassion fatigue, 103
Concepción dam, 54
coronavirus pandemic, 91, 136
corruption: in Guatemala, 192n166; in Honduras, 57, 63, 80–81; in Nicaragua, 185n101; in Puerto Rico, 133–34. *See also* criminal activity; foreign debt
Costa Rica, 48, 62, 69, 70, 77, 85, 185n98
COVID-19 pandemic, 91, 136
criminal activity: deportation and, 80, 189n140, 190n148, 192n167; drug trafficking, 49, 76, 80, 192n166; in Honduras, 80–81, 87–89; human trafficking, 3, 67, 78, 80, 86; in Puerto Rico, 104. *See also* corruption; gangs
crop disease, 141, 145
Cuba, 31, 36, 57, 63, 73, 125, 181n55, 208n93
Cuban migrants, 73, 222n30
Cyclone Idai (2019), 221n19

dams, 54, 88, 96, 199n16
Danish West Indies, 124, 212n127. *See also* U.S. Virgin Islands (USVI)
deaths. *See* casualties

debris removal, 40, 61, 96, 100, 114
debt relief programs, 60, 62–64, 130, 141, 184n82. *See also* foreign debt; public debt
Deferred Enforced Departure (DED), 10, 12, 37, 75, 146, 163n52. *See also* TPS (Temporary Protected Status)
deforestation, 1, 4, 51, 87, 90, 152n7, 219n15. *See also* soil erosion
Denmark, 124
deportations: of Central American migrants, 12, 60, 71–78, 82–83; criminal activity and, 80, 189n140, 190n148, 192n167; by Mexico, 69; of Montserrat migrants, 36, 37; by Trump, 213n136. *See also* detention of migrants; repatriation
detention of migrants, 71, 72, 74, 76–77, 213n136. *See also* deportations
DFID. *See* UK Department for International Development (DFID)
Dinant Corporation, 88–89
disaster-driven migration: from Central America, 69–79; international economic policies and, 15, 63–65, 68, 219n9; landscape transformation by, 219n15; from Montserrat, 14–15, 25–29, 31–37; from Puerto Rico and USVI, 16–17, 130–33. *See also* environmental migrant, as category; war-driven migration
Disaster Housing Assistance Program (FEMA), 128
Disaster Relief Fund (FEMA, U.S.), 16, 108
disaster tourism, 42–43. *See also* tourism
disease, 2, 3, 20, 28, 52, 55, 59, 65, 86, 99, 101, 141, 219n15
displacement statistics, 7, 157n27
Diversity Visa Lottery (U.S.), 146–47
Dole Foods, 55
Dominica, 38, 94, 197n1, 198n2
Dominican Republic, 38, 61, 185n98, 198n2
drinking water. *See* water supply
drought, 43, 48, 49, 85, 86, 108, 141, 142, 221n20
drug trafficking, 49, 76, 80, 192n166. *See also* criminal activity
Duany, Jorge, 118, 120
Duke, Elaine, 111, 114
Duncan, Tim, 106, 205n59

227

INDEX

duties, 7–8, 45, 116, 134, 158n32. *See also* citizenship

earthquakes, 161n39; in Azores, 9, 161n41, 162n45; in Colombia, 61, 183n72; in El Salvador, 12, 193n179; in Haiti, 46; in Nicaragua, 48; in Puerto Rico, 135–36
eco-apartheid, 143–44
economic impact: of hurricanes in Caribbean, 16–17; of Puerto Rican industries on U.S., 120; of remittances, 14, 16, 31, 40, 48, 71, 74, 76, 78, 146, 164n64, 191n160; of U.S. foreign policy in Central America, 58, 81; of volcanic eruptions on Montserrat, 23, 28, 29, 36, 38, 40–43
economic migrants, as category, 49
economic policies, global *vs.* local, 15, 63–65, 68, 219n9
electricity: in Montserrat, 20; in Puerto Rico, 94, 98, 99, 105, 110, 116, 126, 216n164; in USVI, 94, 99
El Salvador, 12, 49, 58, 66, 69, 70, 75, 145, 183n67, 193n179. *See also* Salvadoran migrants and refugees
energy industry, 43, 136
Enhanced Border Control Operational Program, 74
ENSO (El Niño–Southern Oscillation), 48, 176n4
environmentally displaced person, as term, 6, 155n20
environmental migrant, as category, 6, 152n7, 155n20. *See also* disaster-driven migration
Environmental Protection Agency (EPA), 152n3, 200n20
environmental refugee, as term, 155n19
Espaillat, Adriano, 128
Espinoza, Noem de, 57
Estrella Martínez, Luis F., 118
Ethiopia, 221n20
Export Services Act (Puerto Rico), 210n115

Facussé Barjum, Miguel, 88–89, 185n100

Faial, Azores, 9, 10
family reunification, 33
famine, 221n20. *See also* food insecurity
Farenthold, Blake, 115
Federal Reserve Bank (U.S.), 126
FEMA (Federal Emergency Management Agency; U.S.): on 2017 hurricanes, 16–17, 116; denial of housing aid by, 128, 213n138; disaster aid by, 96–102, 106–13, 132–33, 205n63; federal funding of, 16; inter-agency support of, 94–95; total assistance statistics of, 206n70
First Organic Act (1900), 117
fishing industry, 53, 85, 87
flooding: in Central America, 48, 52, 54–55, 90; in Mozambique, 221n19
Flores administration, 56, 57, 65, 68
Florida, 107, 110, 131, 132–33, 197n2, 216n161; Nicaraguan and Cuban nationals in, 73
food insecurity, 48, 141. *See also* famine; poverty
Foraker Act (1900), 117
forced sterilization, 119, 209n106
foreign debt: in Honduras, 55, 62–64, 91; of Nicaragua, 55, 63. *See also* corruption; debt relief programs; public debt
Frade, Leopoldo, 56
France, 63

gangs, 80, 192n167. *See also* criminal activity
geothermal power, 43, 136
Germany, 67, 124
global warming. *See* climate change
Global Witness, 89, 196n205
González, Sonia, 79–80
González-Colón, Jenniffer, 98, 107, 113
Gore, Tipper, 60, 182n66
Green, Al, 127
Grenada, 22, 145
Grenadines, 145
Guajataca dam, 96
Guatemala, 12, 66–67, 84, 145, 183n67, 191n160
Guatemalan migrants and refugees, 72, 73, 75, 187n120

INDEX

Guiding Principles on Internal Displacement (UNHCR; 1998), 8, 157n29, 160n37
Guterres, António, 8, 93, 142, 159n35

Haiti, 12, 38, 46, 61, 84
Haitian refugees, 36, 46
Hall, Nina, 159n35
Hart-Celler Act. *See* U.S. Immigration and Nationality Act (1965)
Hasina, Sheikh, 157n26
Hayes, Ben, 143
health care. *See* medical care
homelessness: in Colombia, 183n72; in Honduras, 56; in Montserrat, 23; in Nicaragua, 48, 49, 51, 69; in Puerto Rico, 135. *See also* housing damage and assistance; poverty
Honduran Christian Council for Development, 57
Honduran migrants and refugees, 15–16, 68, 72–75, 77–84, 189n136
Honduras: disaster aid in, 59–68, 181n55, 197n207; effects of Hurricane Mitch in, 12, 15, 49, 53–59, 84–86; human rights violations in, 87–90; as "post-conflict society," 145; as qualifier of TPS in U.S., 12; violence in, 58–59, 80–81, 88, 196n203
Honoré, Russel L., 109
hospitals. *See* medical care
housing damage and assistance: in Montserrat, 23, 24–25, 26, 27, 31, 33, 34, 38, 40; in Nicaragua and Honduras, 65, 81, 86, 178n18; in Puerto Rico, 96, 120, 124, 125, 128, 131; in USVI, 99–100, 101, 103–4. *See also* homelessness
Housing Victims of Major Disasters Act (proposed), 128
H. R. 1726 (U.S.), 36
"humanitarian stays," 164n61
human rights violations, 87–90
human trafficking, 3, 67, 78, 80, 86. *See also* criminal activity
Hurricane César (1996), 69
Hurricane Eta (2020), 90–91
Hurricane Fifi (1974), 58
Hurricane Georges (1998), 61, 127
Hurricane Harvey (2017), 95, 107, 112, 216n161
Hurricane Hugo (1989), 23, 127, 205n59
Hurricane Iota (2020), 90–91
Hurricane Irma (2017), 16–17, 93–96, 107, 197n2
Hurricane Jose (2017), 108
Hurricane Katrina (2005), 77, 109, 111, 113, 126, 148
Hurricane Maria (2017), 16–17, 93–96, 198n2. *See also* Puerto Rico; U.S. Virgin Islands (USVI)
Hurricane Mitch (1998), 12, 15, 49–59, 69, 84–86
Hurricane Nate (2017), 108
hurricanes: in Caribbean, 177n8; in Central America, 12, 47–48, 84–85, 86–87, 90, 145; effects on migration due to, 142; FEMA's report on, 108–9; in Montserrat, 26, 40, 45, 46; ocean surface temperatures and, 152n6. *See also names of specific hurricanes*
Hurricane Sandy (2012), 113, 148
Hurricane Stan (2005), 12

IIRIRA (Illegal Immigration Reform and Immigrant Responsibility Act; 1996), 37, 71–72, 75
Immigration and Protection Tribunal (New Zealand), 139–40, 217n1
Independent Commission for Aid Impact, 174n104
Individual Investors Act (Puerto Rico), 210n115
Insular Cases, 117, 118
Inter-Agency Standing Committee (IASC), 158n30
Inter-American Development Bank, 62, 63, 64
Intergovernmental Panel on Climate Change (IPCC), 7, 141, 152n6, 158n31
Internal Displacement Monitoring Center, 7, 157n29
International Monetary Fund, 63
International Organization for Migration (IOM), 7, 45, 78, 155n20, 156n24

INDEX

International Red Cross and Red Crescent, 7
Isle de Jean Charles, Louisiana, 148, 223n36

Johnson, Lyndon, 66, 185n97
Johnson, Violet Showers, 31
Jones Act (1917), 118, 119
Jones Act (1920), 114, 116

Kagan, Elena, 80
Kaiser Family Foundation, 101, 129
Keleher, Julia, 134
Kennedy, Edward, 37
Kerry, John, 37
Kiribati, 139–40
Klein, Naomi, 123

Laguna del Pescado dam and reservoir, 54
land mines, 52, 55, 61
landslides, 51, 52, 54–55, 90, 135. *See also* avalanche; coastal erosion
language and migration, 5, 132, 142
Lawyers' Committee for Human Rights, 164n60
Leahy Amendment (1997), 89
LeBrón, Marisol, 117, 123
legal definition of "refugee", 5–6, 9–10, 74, 139–40, 154n16, 161n40. *See also* citizenship
Liberia, 163n52
literacy, 195n195
logging industry. *See* deforestation
Long, Brock, 96, 108
Louisiana, 77, 147–48, 190n151, 205n63, 223n36. *See also* New Orleans, Louisiana
Lynch, Stephen F., 36

Mack, Doug, 116–17
malaria, 52, 59
Mapp, Kenneth E., 99, 107, 112, 113, 114–15, 126
maquiladoras, 56, 64, 78, 184n87
March for Life/Marcha por la Vida protest (2004), 87–88, 196n202
Markey, Edward, 147
Martin, Ricky, 132

Martínez-Otero, Heriberto, 123
Martinique, 22
Mattis, James, 113
McCarran-Walter Act (1952), 31, 170n60, 222n29
McNeill, John R., 2
medical care: drug industry and disasters, 99, 201n26; for Hondurans, 55, 63; for Montserratians, 23, 25, 30, 32; in Puerto Rico, 97, 99, 104–5, 116, 135; in USVI, 95, 100–101, 127, 201n32
medical experimentation, 119. *See also* forced sterilization
Memphis Grizzlies (NBA team), 106
mental health, 101, 129. *See also* medical care; post-traumatic stress
Mexican migrants and refugees, 49, 82, 142
Mexico, 68, 69, 142, 188n133; international assistance from, 57. *See also* NAFTA
migrant detention, 71, 72, 74, 76–77, 213n136. *See also* deportations
migrant rights, 7–8. *See also* citizenship; TPS (Temporary Protected Status)
migration. *See* disaster-driven migration; war-driven migration
Migratory Amnesty, 69
military service, 115, 119, 125. *See also* U.S. military
minimum wage, 119, 123, 185n100, 213n137
mining, 41, 43, 88–89
Mira Loma Detention Center, U.S., 76–77
Miranda, Lin-Manuel, 105, 211n117
Montreal Gazette (publication), 38
Montserrat, 14–15; archival heritage of, 20, 165n1, 166n10; citizenship of, 29, 32, 45; climate change and future of, 44–46; commonwealth status and disaster aid in, 14, 32, 45, 144–45; initial recovery and aid for residents of, 24–31; long-term recovery and conditions on, 37–44, 174n104; migrants with TPS in U.S., 12, 33–37, 46, 172n77; migration tradition of, 31, 166n6; political history of, 20–22, 23–24, 166n6; population statistics of, 14, 31, 33, 41, 43, 170n55; volcanic eruptions in, 14, 19–20, 22–23, 25, 30, 168n24

INDEX

Montserrat Disaster Management Coordinating Agency, 44
Montserratian migrants and refugees, 14–15, 25–29, 31–37
Montserrat Immigration Fairness Act (2004), 37
Montserrat People's Progressive Alliance, 174n98
Montserrat Sustainable Development Plan 2008–2020 (SDP), 40–41
Montserrat Volcano Observatory, 24, 30–31, 42, 43, 167n22
Mozambique, 221n19
mudslides, 51, 52, 54–55. *See also* landslides
Murillo Vargas, Josefina, 52

NACARA (Nicaraguan Adjustment and Central American Relief Act), 73, 188n127
NAFTA (North American Free Trade Agreement), 64
Nansen Principles, 8
NASA (National Aeronautics and Space Administration), 85, 151n1
National Association of Hispanic Journalists, 103
National Climatic Data Center, 49
National Council of Popular and Indigenous Organizations of Honduras, 88
National Disaster Resilience Competition, 148, 223n36
National Oceanic and Atmospheric Administration, 108, 151n1, 194n187
national security, 58, 71, 142–44. *See also* wars
natural calamity provision, 9, 10
Nelson, Bill, 113
Nepal, 84
Nevis, 27, 29
Newland, Kathleen, 142, 144
New Orleans, Louisiana, 77, 147–48, 190n151, 223n35
New York Times (publication), 70, 108
New Zealand, 139–41, 217n1
Nicaragua: civil wars in, 52, 58; disaster aid in, 66, 67, 181n55, 182n64, 183n67, 183n70; effects of Hurricane Mitch in, 12, 15, 49–53, 86, 178n18; IIRIRA and, 71–72; as "post-conflict society," 145; as qualifier of TPS in U.S., 12
Nicaraguan Adjustment and Central American Relief Act (NACARA), 73
Nicaraguan migrants and refugees, 15–16, 71–77, 81–84, 187n120
Nicaraguan Telecommunications Company, 52
Nielsen, Kirstjen, 83
Nixon, Rob, 4
non-refoulement, 8
Norton, Gill, 19
Norway, 8
Norwegian Refugee Council, 153n7

Obama administration, 82, 83, 122, 123, 148, 223n36, 224n39
Obando y Bravo, Miguel, 47
Ocampo González, Bernardo, 52
ocean surface temperatures, 152n6
Oliva, Bertha, 88
Oliver-Smith, Anthony, 139
Organic Act (1936), 125
Organization of African Unity, 5, 154n18
Organization of American States, 52, 154n18
Organization of Eastern Caribbean States, 40, 43
Osborne, Bertrand, 27
Osborne, John, 34–35
Owens, Major, 36
Oxfam, 63

Paleolithic era, 2
Palmer, Larry Leon, 82
palm oil industry, 88–89
Panama Canal, 117
Parenti, Christian, 143
Patriot Ledger (publication), 35
Pattullo, Polly, 23, 39, 168n24, 174n98
Peace Corps, 61
Pence, Mike, 107, 113
Peru, 161n39
Pesquera, Hector, 94

INDEX

pharmaceutical drug industry, 99, 201n26. *See also* medical care
Philippines, 113
Philpott, Stuart B., 31
Physical Development Plan for North Montserrat, 2012–2022 (PDP), 40, 41
plantain industry, 98. *See also* agricultural production
Plaskett, Stacey, 113, 114–15, 126–27, 128, 208n88
post-traumatic stress, 39–40, 101, 129. *See also* mental health
poverty, 81, 86, 122, 125, 129, 214n142. *See also* food insecurity; homelessness
#PRActívate, 102
prehistoric climate change, 2
La Prensa (publication), 81
PREPA. *See* Puerto Rico Electric and Power Authority (PREPA)
protests, 68, 87. *See also* citizen activism
public debt: of Puerto Rico, 110, 121–24, 127, 134, 211n117, 211n125; of USVI, 124, 125–26. *See also* debt relief programs; foreign debt
Public Law 85-892 (U.S.), 9
Puerto Rican migrants, 16–17, 121, 122, 130–33
Puerto Rico: 2017 hurricanes in, 16–17, 93–96, 198n2; calls for statehood of, 133; communication of damage in, 102–6; disaster aid and recovery in, 96–102, 106–13, 128, 133; earthquakes in, 135–36; economy and industries of, 120, 121–22, 126, 211n125, 213n138; political scandal in, 133–35; public debt of, 110, 121–26, 127, 134, 211n117, 211n125; territorial status and disaster aid in, 97, 101, 113–21, 124, 127, 129, 144–45, 210n109, 210n112
Puerto Rico and Virgin Islands Equitable Rebuild Act (proposed), 128
Puerto Rico Electric and Power Authority (PREPA), 98, 133–34, 216n164
Puerto Rico Health Insurance Administration, 134
Puerto Rico Oversight, Management, and Economic Stability Act (2016), 123

Puerto Rico v. Franklin California Tax-Free Trust et al., 123

R2P principle, 6, 156n22
Rabin, John, 116
racism, 36, 104, 121, 127
radio stations: in Montserrat, 24, 26, 43, 168n33; in Puerto Rico, 102, 202n36; in USVI, 101, 104–5. *See also* communication, post-disaster
Ramos, Efrén Rivera, 118
Ramos et al. v. Nielsen et al., 84
Raudales, William Handal, 47
REDD+ (Reducing Emissions from Deforestation and Forest Degradation program), 90, 197n209
refugee: Behrman on term, 156n23; legal definitions of, 5–6, 9–10, 74, 139–40, 154n16, 161n40. *See also names of specific regional groups*
Refugee Convention (UN), 5, 8, 140, 154n16, 159n34
Refugees International, 129
Reichman, Daniel, 78
remittances, 14, 16, 31, 40, 48, 71, 74, 76, 78, 146, 164n64, 191n160
Reno, Janet, 12, 34, 75
repatriation, 13. *See also* deportations
resilience, defined, 3
Reyes, Carlos Arturo "Oscar," 87–88
Robert T. Stafford Disaster Relief and Emergency Assistance Act (1974), 199n12
Rodgers, Malcolm, 55, 86
Rodríguez Cotto, Sandra D., 99, 121
Rodríguez, Miguel Ángel, 69
Rodríguez Rivera, Luis E., 124
Roosevelt Roads naval base, Puerto Rico, 119, 120
Rosselló, Ricardo, 94, *112*, 115, 116, 130, 134, *135*
Rubio, Marco, 113

Salvadoran migrants and refugees, 12, 73, 75, 79–80, 83, 187n120. *See also* El Salvador
Sánchez, José, 79–80

INDEX

Sanchez v. Mayorkas, 79–80
Sanders, Bernie, 128
schools, 127, 130, 134, 183n70
Schumer, Charles, 37
Scott, Rick, 131
sea level rise, 147–48
Seda-Irizarry, Ian, 123
Serrano, José, 113
SERVIR project, 85
sewage management, 53, 55, 59, 98
shock doctrine, 123
Short, Claire, 27
Shotte, Gertrude, 31
shrimp farming, 53, 56
Skelton, Tracey, 25, 30, 40, 168n33
slow violence, 4
Smith, Adam, 113
Smith, Lamar, 13
smuggling. *See* drug trafficking; human trafficking
social media, 102–4, 106–7, 110, 127, 131, 205n62
soil erosion, 87. *See also* coastal erosion; deforestation; landslides
solar power, 105, 136, 204n54
South Tarawa, Kiribati, 139–41
St. Croix, USVI, 95, 97, 125, 212n127. *See also* U.S. Virgin Islands (USVI)
St. John, USVI, 95, 125, 212n127. *See also* U.S. Virgin Islands (USVI)
St. Kitts-Nevis, 29, 32
St. Lucia, 145
St. Thomas, USVI, 95, 97, 125, 212n127. *See also* U.S. Virgin Islands (USVI)
St. Vincent, 22, 145
Sudan, 84
sudden-onset disasters, 4
sugar cane industry, 118
suicide, 101. *See also* mental health
sustainability, defined, 3
Sweden, 67

Taiwan, 181n55
Tamayo, José Andrés, 87
taxation, 121, 122–23, 126
Teitiota, Ioane, 139–41
Telemundo, 102

Temporary Protected Status. *See* TPS (Temporary Protected Status)
Territorial Emergency Management Agency (USVI), 99–100
territorial status and disaster aid in Puerto Rico and USVI, 97, 101, 113–21, 124, 129, 144–45, 210n109, 210n112. *See also* commonwealth status and disaster aid in Montserrat; Puerto Rico; U.S. Virgin Islands (USVI)
Texas, 95, 107, 112, 205n63, 216n161
textile industry, 56, 64–65, 184n89. *See also* maquiladoras
Toronto Star (publication), 22–23
Torres Calderón, Manuel, 78, 91, 183n79
Torruella, Juan R., 118
tourism: in Florida, 131; on Montserrat, 42–43; in USVI, 100, 102, 124, 125
TPS (Temporary Protected Status), 10–13, 146; of Central American migrants, 74–75, 79–84, 189n136; of Haitian refugees, 46; limitations of, 15–16; of Montserratian refugees, 12, 33–37, 46, 172n77. *See also* citizenship; Deferred Enforced Departure (DED); United States
trafficking. *See* drug trafficking; human trafficking
Transitional Sheltering Assistance (FEMA), 128
Transparency International, 63
Treaty of Paris, 117
Trías Monge, José, 120
Truman administration, 119
Trump administration: 2017 hurricane disasters and, 95, 96, 106–16, 127–28, 133; ICE funding by, 213n136; TPS and deportations by, 82, 83, 84
tsunamis, 25, 109, 136, 217n168
Typhoon Haiyan (2013), 113

UK Department for International Development (DFID), 24, 25–26, 28–29, 40, 43, 168n29, 169n40
UK Overseas Private Investment Corporation, 64
Union of Concerned Scientists, 148

INDEX

United Fruit Company (formerly Chiquita Brands), 31, 55

United Nations: on climate change, 141–43; Development Program, 40; Honduran aid by, 57; Human Rights Committee, 140; and Hurricane Mitch, 53; in Montserrat, 21, 40; Office for the Coordination of Humanitarian Affairs, 57; and Puerto Rico, 119; *Refugee Convention*, 5, 8, 140, 154n16, 159n34; Sustainability Development Goals, 90; UNHCR, 6, 8, 157n29, 159n35, 160n37

United States, 15–16; 2017 total disaster damages in, 108; disaster aid as foreign policy of, 59–68, 197n207; disaster-driven migration to, 33–37, 46, 69–79, 171n69, 171n71, 172n77; humanitarian and relief aid by, 15, 58, 96–97; response to climate-driven migration, 8–14. *See also* Puerto Rico; TPS (Temporary Protected Status); U.S. Virgin Islands (USVI); *and names of specific agencies*

Univision, 102

USAID (United States Agency for International Development), 57–58, 59–68, 182n64, 183n67

U.S. Bureau of Insular Affairs, 118

U.S. Citizenship and Immigration Services, 35, 146, 191n162

U.S. Congressional Black Caucus, 36

U.S. Department of Defense, 96–97, 98, 109, 116

U.S. Department of Homeland Security, 11, 34, 79, 96

U.S. Department of Justice, 33, 75, 78, 146

U.S. Disaster Relief Act (1974), 199n12

U.S. Immigration Act (1952), 31, 170n60, 222n29

U.S. Immigration Act (1990), 146–47

U.S. Immigration and Customs Enforcement (ICE), 213n136

U.S. Immigration and Nationality Act (1965), 9–10, 161n42

U.S. Immigration and Naturalization Service (INS), 10, 72, 74

U.S. military: effects of climate change on, 220n18; in Honduras, 58, 59, 61; in Puerto Rico and USVI, 94–95, 96–97, 109, 119, 124; service members of, 115, 119, 125

U.S. Naval Construction Battalion, 61

U.S. Northern Command, 94–95, 107–8, 198n7. *See also* U.S. military

U.S. Refugee Act (1980), 9, 10, 146

U.S. Refugee Admissions Program, 146, 164n62

U.S. Refugee Protection Act (2010), 164n62

U.S. Refugee Relief Act (1953), 9

U.S. Southern Command, 182n63. *See also* U.S. military

U.S. Virgin Islands (USVI): 2017 hurricanes in, 16–17, 93–96, 198n2; communication of damage in, 102–6; disaster aid and recovery in, 96–102, 106–13, 133, 208n88; economy and industries of, 124, 125, 126; population statistics in, 215n157; public debt of, 124, 125–26; territorial history of, 124–25, 212n127; territorial status and disaster aid in, 97, 101, 113–21, 124, 127, 129, 144–45, 210n109, 210n112

Vázquez Garced, Wanda, 135, 136

Vega, Christian Sobrino, 134

Velázquez, Nydia, 113–14, 147

Virgin Island migrants, 16–17, 130, 132. *See also* U.S. Virgin Islands (USVI)

Virgin Island National Park, St. John, USVI, 125

Virgin Islands Daily News (publication), 104

volcanic eruptions: in Azores, 9, 161n41, 162n45; disaster tourism and, 42–43; in El Salvador, 193n179; in Martinique, 22; in Montserrat, 14, 19–20, 22–23, 25, 30; in Nicaragua, 48; in St. Vincent, 22

voting rights, 119, 121, 125, 132–33. *See also* citizenship

Vox (website), 133

vulnerable populations, 3–4

vulture capitalism, 103, 122, 123

WAPA Radio, 102, 202n36
war-driven migration, 68–69, 219n16, 221n19. *See also* disaster-driven migration; *and names of specific nations*
War of 1898, 116, 117
Warren, Elizabeth, 128
wars: in Ethiopia, 221n20; in Guatemala, 66–67; in Honduras, 58–59, 80–81, 88, 196n203; in Nicaragua, 52, 58, 182n63. *See also* casualties; national security
Washington Post (publication), 35, 129
Washington Times (publication), 83
Water Island, USVI, 95, 124. *See also* U.S. Virgin Islands (USVI)
water pollution, 87
water supply: in Central America, 53, 55, 61, 145; in Kiribati, 139; in Montserrat, 20, 30; in Puerto Rico and USVI, 94, 97, 98, 100, 103, 116

weather, defined, 151n1
welfare programs, 26, 120, 125, 128, 146
White, Edward Douglass, 118
Whitefish Energy, 133–34
wildfires, 2, 108, 152n3
wind power, 136
women's reproductive health, 119, 209n106
women's work, 56, 64, 78, 129, 184n87
World Bank, 40, 62–63, 142, 143, 160n37
World Central Kitchen, 105

Youth Development Institute (Puerto Rico), 129
Yulín Cruz, Carmen, 110–11, 112

Zelaya, Manuel, 67, 88, 89, 185n100
Zeledón, Felicitas, 51